Practical Work in Science Education: Recent Research Studies

Practical Work in Science Education: Recent Research Studies

Editors: John Leach and Albert Paulsen

Roskilde University Press

John Leach and Albert Paulsen (eds.)
Practical Work in Science Education

1.edition 1999 Roskilde University Press

© Roskilde University Press

Cover: Torben Lundsted
Typeset: Narayana Press, Gylling, Denmark
Print: Narayana Press, Gylling, Denmark

ISBN 87-7867-079-9

Published by:

Roskilde University Press
Rosenoerns Allé 9-11
DK-1970 Frederiksberg C
Denmark
Phone (+45) 35 35 63 66
Fax: (+45) 35 35 78 22
E-mail: slforlag@sl.cbs.dk
www.samfundslitteratur.dk

Kluwer Academic Publishers
P.O. Box 17
3300 AA Dordrecht
Holland
Phone: (+31) 78 639 23 92
Fax: (+31) 78 639 22 54
E-mail: services@wkap.nl

All rights reserved.
No parts of this book may be reproduced or transmitted in any form or by any means, electronic or mechanical, including photocopying, recording, or by any information storage or retrieval system, without permission in writing from the publisher.

Contents

Introduction, *John Leach and Albert Paulsen*   7

Section 1:
Aims and rationales for practical work   17

Introduction, *Albert Paulsen*   17
Practical work in School Science – some questions to be answered, *E.W. Jenkins*   19
"Mapping" the domain – varieties of practical work, *R.H. Millar, Jean-François Le Maréchal and Andrée Tiberghien*   33
Envisionment in Practical Work. Helping pupils to imagine concepts while carrying out experiments, *J. Solomon*   60
TIMSS Performance Assessment – a cross national comparison of practical work, *P.M. Kind*   75
Touched by a disgusting fish. Dissecting squid in biology lessons in a comprehensive school. *P. Szybek*   96

Section 2:
Practical work and learning about science   115

Introduction, *John Leach*   115
A marriage of inconvenience? School science practical work and the nature of science. *G.M. Ntombela*   118
Learning science in the laboratory. The importance of epistemological understanding. *J. Leach*   134
The interaction between teaching styles and pupil autonomy in practical science investigations – a case study *R. Watson, J. Swain and C. McRobbie*   148
Is *authentic* appropriate? The use of work contexts in science practical activity. *S. Molyneux-Hodgson, R. Sutherland and A. Butterfield*   160

Section 3:
Practical work and teaching scientific concepts 175

Introduction, *John Leach* 175
Labwork activity and learning physics – an approach based on modelling. *A. Tiberghien* 176
Modelling Student's cognitive activity during resolution of problems based on experimental facts in chemical education. *J-F. Le Maréchal* 195
A challenge for lifelong science understanding. The role of "lab work" in primary school science. *M. Gagliardi, N. Grimellini-Tomasini and B. Pecori* 210
Longitudinal study on lab work and 10-12-year-olds' development of the concepts of transformation of matter *O. Eksilsson* 229

Section 4:
Practical work outside the laboratory 247

Introduction, *Albert Paulsen* 247
The theoretical and practical aspects of actively involved enquiry learning in a controlled environment (the Educational Greenhouse). A case study based analysis. *M. Dvir and D. Chen* 249
Changing teachers' practise. Practical work in environmental education. *P. van Marion* 264

Section 5:
Models of student cognition in practical work: Perspectives from one research programme 279

Introduction, *Albert Paulsen* 279
The influence of students' individual experiences of physics learning environments on cognitive processes. *C. von Aufschnaiter, A. Schoster and S. von Aufschnaiter*
The influence of learning environments on cognitive processes. *A. Schoster and S. von Aufschnaiter* 297
How to interact with students? The role of teachers in a learning situation. *M. Welzel, C. von Aufschnaiter and A. Schoster* 313

Contact details for first authors of chapters in this book 328

Practical Work in Science Education: Recent Research Studies

Introduction: How this book came about, and its contribution to the literature

John Leach and Albert Paulsen

Science enjoys an increasingly prominent place in the school curriculum of many countries. For example, in a 1992 report addressing schooling in modern European society, Husén and his colleagues argue for science education to support future specialists, those who may encounter science in their future work, and the needs of citizens living in modern societies. The American Association for the Advancement of Science make a similar case (AAAS, 1989).

However, there are considerable variations in the provision and traditions of school science education across Europe (see Tiberghien et al., 1998). For example, although there is a formal requirement to teach science in the primary curriculum of most European countries, there are considerable variations in the amount of primary science taught. Another major difference in the culture and traditions of school science education between European countries is in the use of practical work as a teaching method. In some countries, notably the U.K., France and the Nordic countries, practical work is commonly used as a teaching method in school science. By contrast, practical work is extremely rare in school science in Italy and Greece. It is interesting that much literature coming out of countries with a strong tradition of school science practical work concentrates on questioning the limitations of practical work as a vehicle for a number of ambitious aims (e.g. Woolnough and Allsop, 1985; Hodson, 1996), whereas there are moves afoot in Italy and Greece to increase the use of practical work as a teaching method in school science.

We hope that the papers in this book present a balanced view of practical work, both outlining possibilities and raising difficulties. There is already a substantial literature addressing practical work on science education, including books (e.g. Hegarty-Hazel, 1990; Hodson, 1998; Solomon, 1980; Wellington, 1998; Woolnough and Allsop, 1985), research project reports (e.g. APU, 1984; Séré et al., 1998), research papers (e.g. Tamir et al., 1992; Lunetta and Tamir, 1979) and reviews (e.g. Lunetta, 1998). This corpus of work tends to have come out of countries with a strong tradition of practical work in science education. In spite of a wide range of teaching strategies for practical work in these countries, very little is known about the kind of learning that is produced by different teaching strategies. It is generally assumed that practical work contributes substantially to conceptual learning and to the attainment of skills associated with 'the methods of science'. Although countries with a long tradition of practical work assess students' conceptual knowledge, very few actually assess students' practical skills. Although some assessment systems, such as the Danish oral examination for the leaving certificate, includes practical work, no benchmarks for these assessments are given. In the U.K., the APU (Assessment of Performance Unit) presented an extensive framework for assessing U.K. students' performance in practical work (APU, 1984), and this has been drawn upon in the UK and elsewhere in designing assessment systems for practical work. This framework was based upon a view of science as 'an experimental subject concerned fundamentally with problem-solving.'(p.4). Six categories of science performance were identified from this view of science, namely

- Use of graphical and symbolic representation
- Use of apparatus and measuring instruments
- Observation
- Interpretation and application
- Planning of investigations
- Performance of investigations

This account of science performance is open to criticism on a number of grounds. It places undue emphasis upon laboratory practices, without considering the influence of existing public scientific knowledge on investigative work. It also represents an empiricist view of the generation of scienti-

Introduction

fic knowledge, in which knowledge claims are presented as being generated in an unproblematic way from data. Perhaps more fundamentally, this framework for school students' science performance appears rather naive in its portrayal of the purposes of school science practical work, by its failure to recognise that the vast majority of practical work in schools is carried out to illustrate *existing* scientific knowledge rather than to generate new knowledge claims. These issues are addressed by Millar and Driver (1987) and Hodson (1996). In spite of these difficulties, however, the influence of the APU framework can be seen in much research on school science practical work carried out during the late 1980s and 1990s.

In addition, there is a body of research mapping out the learning objectives that teachers attribute to practical work in science teaching (e.g. Boud, 1973; Kerr, 1964; Welzel et al., 1998). In response to survey questions, teachers generally state that the main aims of practical work are to help students to link theory to practical situations, to learn about experimental methods, and to learn laboratory techniques.

However, much less attention has been given in the literature to the relationship between instructional activities and student learning during practical work. The European Commission funded the international research project *Labwork in Science Education* between 1996 and 1998, co-ordinated by Marie-Geneviève Séré. Several products from the LSE project are reported in this book. An initial task was to develop a more comprehensive conceptualisation of the aims of practical work, and how these aims relate to the design of teaching activities that students are ask to undertake. The resulting *Map of Labwork* is described by Robin Millar, Jean-François Le Maréchal and Andrée Tiberghien in this book. A key aim of the project was to conduct a number of case studies in order to examine critically how the design and context of practical work teaching activities influences what students actually do during practical work, and what they learn from practical work (Séré et al., 1998). Findings from several case studies of this kind are reported in this book, coming out of the LSE project and elsewhere (e.g. Le Maréchal, Tiberghien, Molyneux-Hodgson and colleagues, and the papers included in Section 5 of the book).

In our view, such papers have the potential to make an important contribution to the literature. There are now a number of studies which provide compelling evidence that some instructional strategies are signifiy more effective at promoting students' understanding of specific conceptual learning goals, than others (e.g. Brown and Clement, 1991; Viennot, 1998). It would be useful to see a range of similar studies being reported, which evaluate the effectiveness of teaching strategies against some of the learning goals identified for practical work. Such studies would provide invaluable empirical information to inform debates about the cost-effectiveness of practical work, and the amount of time that ought to be devoted to practical work compared to other teaching methods.

The papers in this book were originally presented at the conference *Practical Work in Science Education: The Face of Science in Schools*, held at the Royal Danish School of Educational Studies, Copenhagen, in May 1998. Participants from 18 countries attended the conference. Although most participants at the conference were European, researchers from Australia, Israel, Mauritius, South Africa and the USA were present. The 18 papers in this book were selected for publication from amongst the 39 papers submitted for publication because they were judged to be of broad international interest, reflecting something of the diversity of the practice of both school science teaching and science education research in Europe. The following researchers were members of the scientific committee that helped us to review the papers:

Björn Andersson, Gothenberg University, Sweden

Annemarie Møller Anderson, Royal Danish School of Educational Studies, Denmark

Per Morten Kind, Trondheim University, Norway

Kirsten Nielsen, Royal Danish School of Educational Studies, Denmark

Jenny Lewis, The University of Leeds, UK

Helene Sørensen, Royal Danish School of Educational Studies, Denmark

Robin Millar, York University, UK

Poul Thomsen, Århus University, Denmark

Andrée Tiberghien, University of Lyon 2, France

Manuela Welzel, Bremen, University, Germany

We are very grateful to all members of the scientific committee for their hard work and pertinent advice through the reviewing process.

Introduction

This book is organised around 5 sections, and the papers in each section are introduced by one of us. The first section contains five papers which address the aims and rationales for practical work as a teaching method. Each paper makes a distinctive contribution to the existing literature, whether by drawing upon theoretical perspectives not previously used in the context of practical work (e.g. Szybek), by questioning some of the taken-for-granted aspects of school science practical work (e.g. Jenkins), or by developing and refining perspectives in the existing literature (e.g. Millar et al.).

One of the most commonly articulated aims for practical work is to engender amongst students an understanding of how theoretical scientific knowledge is used to explain natural phenomena and inform action. The second section of the book includes papers which address students' learning about the nature and practice of science itself during practical work. Each paper in its own way illustrates the difficulties inherent in specifying what it is *about* science that we believe students ought to learn, and how this might be taught through practical work, in contrast with the 'authentic science' movement in North America (e.g. Roth, 1995).

The third section of the book contains four papers which consider the rationale for using practical work in the teaching of scientific concepts, and the effectiveness of practical teaching at promoting conceptual understanding. As indicated earlier, we see this as a particularly useful direction for future research in science education. The fourth section of the book contains two papers which address practical work outside the conventional environment of the teaching laboratory, raising issues of curriculum focus and teacher expertise.

The final section of the book contains three papers from members of one research group. This group, lead by Stefan von Auschnaiter at the University of Bremen, has been working for some years to identify and characterise the processes by which students learn science during various teaching activities, including practical work. The three related papers address aspects of the group's research programme, including the influence of learning environment and interactions on student cognition.

The contributors to this book come from various linguistic backgrounds. Our goal as editors was to support non-native English speakers in articulating their ideas clearly in English, through dialogue and negotiation, rather than imposing linguistic norms on contributors. We therefore hope that the chapters will be read sympathetically for meaning.

John Leach and Albert Paulsen

References

AAAS (1989) *Project 2061: Benchmarks for Scientific Literacy* Washington DC: AAAS

A.P.U. (1984) *Science Report for Teachers 2: Science assessment framework age 13 and 15* London: Department of Education and Science.

Boud, D.J. (1973) *The laboratory aims questionnaire: a new method for course improvement?* Higher Education, 2, 81-94

Brown, D. and Clement, J. (1991) *Classroom teaching experiments in mechanics* in R. Duit, F. Goldberg and H. Niedderer (Editors): *Research in Physics Learning: Theoretical and Empirical Studies* Kiel, Germany: IPN

Hegarty-Hazel, E. (1990)(Ed.) *The student laboratory and the science curriculum* London: Routledge

Hodson, D. (1996) *Laboratory work and scientific method: three decades of confusion and distortion* Journal of Curriculum Studies, 28, (2), 115-135.

Hodson, D. (1998) *Teaching and learning science: towards a personalised approach* Buckingham, UK: Open University Press

Husén, T., Tuijnman, A. and Halls, W. D. (1992) *Schooling in Modern European Society: A Report of the Academia Europeae*, London: Pergamon.

Kerr, J. F. (1964) *Practical work in school science* Leicester, UK: Leicester University Press

Lunetta, V.N. (1998) *The school science laboratory: historical perspectives and contexts for contemporary teaching* in K. Tobin and B. Frazer (Eds.) *International Handbook of Science Education* Dordrecht, NL: Kluwer

Lunetta, V.N. and Tamir, P. (1979) *Matching laboratory activities with teaching goals* The Science Teacher, 46, (5), 22-24

Millar, R. and Driver, R. (1987) *Beyond Processes* Studies in Science Education, 14, 33-62.

Roth, W-M. (1995) *Authentic School Science: knowing and learning in open-inquiry science laboratories* Dordrecht, NL: Kluwer

Séré, M-G., Leach, J., Niedderer, H., Paulsen, A. C., Psillos, D., Tiberghien, A., Vicentini, M. (1998) *Final report of the project 'Labwork in Science Education' to the European Commission* Orsay, France: Université Paris-Sud XI (available at http://www.physik.uni-bremen.de/physics.education/niedderer/projects/labwork/papers.html)

Solomon, J. (1980) *Teaching science in the laboratory* London: Croom Helm

Tamir, P., Doran, R. L. and Chye, Y. O. (1992) *Practical skills testing in science* Studies in Educational Evaluation, 18, 263-275

Tiberghien, A. and 16 others (1998) *science teaching and labwork practice in several European countries* Working Paper 2, Labwork in Science Education Project.

Lyon, France: Université Lyon 2 (available at http://www.physik.uni-bremen.de/physics.education/niedderer/projects/labwork/papers.html)

Viennot, L. and Rainson, S. (1998) *Design and evaluation of a research-based teaching sequence: the superposition of electric fields* International Journal of Science Education, 14 (4) 475-487

Wellington, J.J. (1998)(Ed.) *Practical work in school science: which way now?* London: Routledge

Welzel, M., Haller, K., Bandiera, M., Hammelev, D., Koumaras, P., Niedderer, H., Paulsen, A., Robinault, K. and von Aufschneiter, S. (1998) *Teachers' objectives for labwork: Research tool and cross-country results* (available at http://www.physik.uni-bremen.de/physics.education/niedderer/projects/labwork/papers.html)

Woolnough, B.E. and Allsop, T. (1985) *Practical work in science* Cambridge: Cambridge University Press

Section 1

Aims and rationales for practical work

Aims and rationales for practical work – introduction

The papers in the first section contribute to comprehensiveness and perspective of the book. Although the first three papers are conceptual rather then reporting on empirical research, they contribute substantially to setting the scene of practical work today.

In the first paper, Edgar Jenkins takes his starting point in the history of practical work in schooling for further studies in Science. While the purpose of schooling has changed from being a preparation for further specialist studies for a small number of the most able students, to an educational enterprise geared to the needs of all future citizens, the methods of teaching practical work may not have changed to meet these fundamental differences in emphasis. Another significant point questioned by Edgar Jenkins is the profound belief among teachers that practical work is a condition for understanding, explaining and for learning science.

Not very much research in practical work is about how to link these beliefs and intentions on the part of teachers with teaching practice and learning. The European research project *Labwork in Science Education* (LSE) made an effort to investigate these links. A classification scheme for practical work was developed as a research tool for this project. Robin Millar, Jean-François Le Maréchal and Andrée Tiberghien, who developed the classification scheme, offer a modified version suitable for the school level. It constitutes a comprehensive tool for analysing and categorising the numerous activities of practical work. For the convenience of the reader the classification scheme is provided in its full length in an appendix. The results of the analyses carried out in the LSE project showed that many activities in practical work rely on routine or even ritual actions. They demand skills and rational knowledge on an instrumental level. Explanations were not asked for before and after the activity.

Explanations and imagination, or as Joan Solomon says in her paper 'explanatory images', are essential for science learning. Pupils should not just register what happens. They need explanations and models and must be helped to make sense of their practical activities. Play may have an important role by creating confidence in science activities and intentions for explorations. In her paper about practical work and envisionment Joan Solomon links playful activities with exploration and explanatory images.

This view of practical work is quite different from the kinds of activities used by the Assessment Performance Unit (APU) in the UK. The APU considered practical work as essentially problem solving activity, where rules have to be followed and routine activities are rewarded. Nevertheless, these kinds of practical activities, as commonly introduced to pupils in the UK, had a 'pay off' in the TIMSS (Third international Mathematics and Science Study) according to Per Morten Kind. His careful study of the TIMSS activities and the analyses and comparison of the results achieved in three selected countries enlightens and substantiates the reasons for the differences observed. Again the feasibility of stereotypes of practical activities is questioned because of the lack of 'explanatory images'.

Very far from the stereotypes of practical work, involving the use of formal skills in systematically experimenting, we find the research report from Piotr Szybek. On the contrary, his phenomenological approach makes a fascinating contrast to any attempt to harness practical activities in formal schemes.

Albert Paulsen

Practical work in School Science
– some questions to be answered

E.W. Jenkins

This chapter reviews some of the questions that have been asked about the role and effectiveness of laboratory work in school science and explores critically some of the answers that have been given. In the light of these answers, an attempt is made to reassess the contribution that practical work can make to pupils' understanding of science as a creative and imaginative activity, rather than as a set of laboratory operations.

Practical work in some form conducted by pupils in specially designed laboratories is now a standard feature of secondary school science teaching in most countries of the world, and where this is not the case, such work is regarded as a clear *desideratum*. In many countries, the notion of practical science has now also secured a place in the science education of younger pupils of primary school age, reflecting national and international commitments to promoting 'science for all' and, in the longer term, a greater understanding of science by the public at large. From one perspective, the centrality now accorded to practical work in school science is hardly surprising. As Joan Solomon has pointed out, science belongs in the laboratory 'as naturally as cooking belongs in a kitchen and gardening in a garden' (Solomon 1980). Looked at from a different perspective, however, a number of questions begin to emerge and it is with some of these questions that this paper is concerned.

I want to bring these questions into focus by commenting briefly upon the terms and conditions upon which laboratory teaching of science came to secure a place in the secondary school curriculum. There are, of course, differences between countries, not least between the United States of America and those countries, including England, that share the European tradition of selective schooling, but, for the purposes of this brief paper, these differences can be ignored. If I use England as my example, the practical teaching of science became an established feature of school science education in the last quarter of the nineteenth century. It was in this period that laboratories for teaching science at school level were first designed and built in significant numbers. It was in this period that laboratory manuals and workbooks were first written, practical examinations devised, and an attempt made to provide practical work with a clear rationale, of which the most familiar and enduring is perhaps the teaching of scientific method, a claim associated with the work of Henry Edward Armstrong, the sesquicentenary of whose birth falls this year. Of critical importance, however, is the fact that the practical teaching of science found a home only within those schools, which enjoyed close historical connections with the universities. In the case of England, it was the public (i.e. independent) and grammar schools, which built the earliest laboratories and taught practical science. It was not until the second half of the twentieth century that other schools, catering for the majority of the population, began to teach practical science on a significant scale and, in the case of primary schools, it was necessary to wait for the legislation of 1988 to secure a place for science in the education of all young children.

The consequences of this are too familiar to require detailed rehearsal. In broad terms, school laboratory work became part of a pre-professional science education, appropriate for those who intended to pursue the study of science beyond school. The activities in which pupils engaged mirrored those, which occupied science undergraduates and the laboratories in which they worked were designed for much the same purpose. The school chemistry laboratories with their fume cupboards, Kipps' apparatus and racks of reagents were replicas of the analytical laboratory, and school physics laboratories with their rigid benches, tangent galvanometers, barometers and other instruments revealed the emphasis placed in nineteenth century physics education upon precise and accurate measurement. Biology, of course, was a different matter. Today, perhaps the most popular of the three sciences with students, biology came very late into the secondary school curriculum, partly, it should be noted, because it was not thought to lend itself to practical work by pupils.(As an indication of the centrality of practical work in science teachers' thinking, this argument is sometimes used today to oppose the inclusion of earth science in the science component of the national curriculum in England and Wales). This serves as an important reminder not only that the curriculum histories of physics, chemistry and biology are significantly different but that, when discussing the merits and claims of laboratory teaching, it may sometimes be appropriate to recognise the distinctions between the various sciences.

A prisoner of history?

My first broad question which I would invite you to consider, therefore, is whether school science teaching is still too much a captive of this historical legacy. Much has, of course, changed. The racks of reagents, the Kipps' apparatus, the tangent galvanometer, the Lees's disc and all that they represent have all gone as chemistry and physics have themselves undergone profound changes. Curricula have been reformed and new science teaching apparatus devised. Perhaps most importantly, more science is being taught to more pupils than at any time in history: mass secondary education is now the norm in industrialised countries. Despite this, science no longer commands neither the authority it once had nor, it seems, the appeal it once held for young people as a career. What changes, if any, are needed to practical science at school to accommodate the profound changes that have taken place in the scientific, social and political context of school science education since laboratory science was first schooled in the late nineteenth century?

Where they have occurred, changes in the patterns of secondary schooling and the development of science within primary education might have been expected to promote an informed and well-grounded debate about the purposes or laboratory teaching. My impression, however, is that the debate has been rather sterile, reflecting beliefs and assumptions rather than any careful evaluation of such evidence as might be available, and rather too readily blurring, even alloying, claims that are essentially educational with others that are philosophical or epistemological. A recognition that investigation lies at the heart of science does not lead directly to discovery learning and an acknowledgement that scientists hypothesise does not, of itself, privilege science over, or demarcate it from, other activities and disciplines. Not surprisingly, therefore, the voluminous literature concerned with practical work is marked by a multiplicity of sometimes conflicting claims and occasionally irreconcilable, even unattainable, goals. Conclusions tend to be asserted, not arrived at. There are long lists of 'aims for practical work', aims, incidentally, that sometimes differ in emphasis or in other ways between teachers and taught, and that are often differentiated when they relate to able and less-able pupils. Some of these aims are essentially cognitive, others affective, dealing with motivation and attitudes, and yet others concerned with the acquisition of technique and manual skill. The language used to describe the virtues of practical work has, of course, changed over the past century or so, mirroring changes in both philosophy and psychology, or, more accurately, educators' responses to them. As far as philosophy is concerned, the positivism and Huxley's 'clear cold logic engine' of the nineteenth century have given way to a much more complex picture, which has led Laudan and others to conclude that we have

> (…) no well confirmed picture of how science works, no theory of science worthy of general assent. We did once have a well developed and historically influential philosophical position, that of positivism or logical empiricism, which has by now been effectively refuted. We have a number of recent theories of science which, while stimulating much interest, have hardly been tested at all. And we have specific hypotheses about various cognitive aspects of science, which are widely discussed but wholly undecided. If any extant position does provide a viable understanding of how science operates, we are far from being able to identify what it is. (Laudan et al. 1986. p. 142).

Ravetz has summarised the position much more succinctly, describing the ideology of science, which has sustained science teachers for generations as 'antique' (Ravetz 1990). His judgement arguably applies with particular force to school laboratory work.

Within psychology, faculty psychology, behaviourism and all that they imply for teaching have yielded to one of the many varieties of constructivism (Phillips 1995) with its much more progressive, although not necessarily more effective, approach to teaching and learning.

What has not changed, however, is the underpinning belief that in some way school science education should seek to replicate the supposed methods of science itself, whether this belief is expressed in terms of 'process' science', 'investigative' science', science as 'enquiry' or in some other terms. My second broad question, therefore, asks whether such replication is feasible and, assuming it is, invites consideration of the extent to which school science education should seek to realise it *through laboratory teaching*.

Replicating science?

I have already hinted at the answer to the first part of this question. Methods are prominent within analyses of science, as one possible means of explaining its success, i.e., its almost unique ability among human activities to engender agreement and demonstrate progress. Science engages with the physical universe by deploying distinctive methods, which ensure, or render more probable, the discovery of true, or at least serviceable, knowledge. The training and systems required to bring this about, to minimise error and help create secure knowledge in the face of human fallibility and prejudice have been well described by Ravetz (Ravetz 1970). However, whether such methods exist, can be described, are indeed employed by practising scientists at the individual level, or are the causes of their success, have been the subject of debate since science was judged sufficiently distinct and progressive to merit some special account. Numerous explorations of the issue have been undertaken from the point of view of the philosophy, history and, latterly, the sociology of science. If these are thought to be too eccentric to science itself, let me quote the Nobel prize winning Peter Medawar. Noting that 'It is not at all usual for scientists to write about the nature of scientific method' and that scientists 'are not in the habit of thinking of matters of methodological policy', he comments that the minority of scientists who have received instruction in scientific methodology 'seem no better off' for having done so (Medawar 1982 73, 80). The message seems clear. Those who would seek to replicate the methods of science when teaching science in school face formidable difficulties. Descriptive headings and algorithmic protocols simply will not do. Professional scientists, or those who actually do experimental work, no doubt undertake activities which could be called 'hypothesising', 'observation' and so on. No doubt they also make use of variable handling schemes of various kinds. But there is nothing about such approaches that is specific to science and thereby explanatory of its power. Characterising the practice of natural science requires an altogether more complex framework, if indeed it can be done. In most academic study of science as an activity, the emphasis has moved away from an attempt to generate universal cognitive foundations, which guarantee success. The work is more descriptive in character, and focuses on such issues as the meaning of the claim that the practices of science are rational, whether the conceptualisations of the physical world generated within scientific practice are uniquely predetermined or underdetermined, in what sense the entities and regularities described within science can be judged as 'real', and whether and to what degree, scientific practice is independent of, or implicated in, wider social and intellectual processes. All of these issues, and many more, are the subject of endless and lively debate. They constitute the field of science studies (Jasanoff et al. 1995). Perhaps the essential point here is that science is now seen to address the material world through diverse, loosely-coupled material and intellectual practices, and the working of these practices is seen to be inescapably socially and institutionally situated. Science is not a mechanistic activity and it cannot, for quite fundamental reasons, be equated with the work of individuals in the laboratory.

It is perhaps time for those who seek to justify laboratory teaching by reference to an outdated philosophy of science to recast their arguments, and thereby, their understanding of the educational functions of the laboratory, in terms of these material and intellectual practices. It is these practices which structure and generate experience in, and of, the natural world, and they are specific to the material, social, technological, historical, political and economic context in which they occur. Practices are complemented by books, apparatus, materials, words and everything else necessary to engage with them. Roth, drawing upon the ideas of, for example, Lave, Rorty and Wittgenstein, explores what this different theoretical perspective might mean in an account of a three year attempt to bring about change in science teaching at an elementary school in Canada (Roth 1998).

Before leaving this issue, I want to make three points. The first is to defend myself against the possible charge that I have exaggerated my case and that, to the extent that laboratory teaching seeks to replicate what scientists do, my concern is misplaced or ill-founded. In my defence, I offer the following description of how the 'cognitive processes' of science are said to operate, drawn from a book published in 1995 and advocating the greater use of investigations similar to those undertaken in schools in England and Wales as part of the national curriculum.

> The facts, skills and understandings can be envisaged as information or patterns in the brain's memory banks ... those 'hard disk stores' will contain ideas about speed, measurement of distance, skill routines about using instruments, notions of a fair-test ...The central processing unit will ... examine the problem and look on the hard disc for help ... These ... must be 'processed', via a series of thought patterns that we label hypothesising, or predicting or whatever ... Hypothesising, or predicting ... are different operations of this central unit ... (Gott and Duggan 1995 p. 27).

My second point relates to science as it is practised towards the end of the twentieth century. There are references in the literature to new systems of knowledge production, sometimes referred to a technoscience but embracing rather more than this term normally implies. The difference is that much, perhaps most, science (in the sense of what scientists actually do) is now integrated with capital, is transdisciplinary in character, and directed towards the generation of knowledge in the context or its application with all that this implies for quality control (Gibbons *et al.* 1994). Redner captured this well when he referred to science changing its ends.

> It is no longer the old science of the last few centuries. That old science is coming to an end in the sense of approaching the limits of its potential scope...Contemporary science is worldly in every sense of the word and quite different in its essential character from the European science of the recent past...these differences are apparent in all dimensions of scientific research, intellectual, instrumental and organisational. They are also revealed in the changed relations of science, technology and production (Redner 1987, p. 15).

My third point is that by promoting an algorithmic view of science, coupled with a firm sense that science is about the certain solution of well-bounded problems, we run the risk of presenting young people with a view of science that does not accord with their own direct or vicarious experience of it. There is now ample work in the public understanding of science, especially that concerned with so called citizen science (e.g., Layton *et al.* 1993; Irwin and Wynne 1996), to confirm that, beyond the world of the laboratory, science is often uncertain, contentious and an inadequate basis for action or decision making. The case of BSE in the United Kingdom is perhaps the most glaring example of the difficulties that arise when science merges with power, whether this be political or financial. How is the laboratory science presented at school to be accommodated to issues such as this? In what ways, for example, does this laboratory science help students understand the notion of risk that is so central to many aspects of the interaction of science, society and technology?

In posing this question, I have gone some way to answering the second part of the broader question I raised a few moments ago about the extent to which the practical teaching of science in the laboratory can help students capture something of the nature of modern science. It seems to me that, in teaching practical science, we may be still too wedded to a narrow, rather than a generous view of the nature of science, one that owes too much to a discredited philosophy at the expense of other insights, that is mechanical and algorithmic rather than imaginative and creative, and which does not accord with the experience of young people as they encounter science in its applications in the world beyond school. If this view is accepted, then the role of the laboratory in teaching science needs to be reassessed and our energies directed towards translating some of my concerns into a curriculum response that is appropriate to the educational needs of students at the end of the millennium. To borrow something from the subtitle of this conference, we need to change the face of science in schools.

The laboratory as a resource for learning

It can, of course, be argued that while the practical teaching of science may seek to mirror what scientists do, its principal concern is enabling, or helping, students to understand scientific ideas. This effectively categorises practical work as a pedagogical strategy, and only one among a number directed towards students' learning. It also points towards my next broad question, namely, how effective is it as a learning device? To begin to answer this question, some preliminary attention needs to be given to what it is students are to learn as a result of their practical work. Having identified this, it then makes sense to turn to the research literature to seek such answers as are available, although it is important to recall my earlier comment that as far as teachers are concerned, the aims of practical work depend upon the age and ability of the children being taught. However, some of these aims are expressed in a form that is simply not testable and when this is not the case, the aims are all too often corruptions of what are really assessment objectives or, to put the matter perhaps more charitably, the outcomes of attempts to reduce practical work to what is measurable. The consequences of this, much in evidence in England and Wales, is the reduction of practical work to a set of techniques or allegedly distinct skills and a consequent frustration of its educational potential.

That potential touches upon a number of different areas which Hodson (1993) has brought together into the following broad categories.

Motivation, by stimulating interest and enjoyment
The acquisition of laboratory skills
Promoting the learning of scientific knowledge
Providing an insight into scientific method and developing expertise in using it
Developing certain 'scientific attitudes', such as open-mindedness, objectivity and a willingness to suspend judgement

A number of comments can be made about this list. First, it is a diverse list that ranges from the promotion of attitudes, through the learning of scientific concepts to the acquisition of motor skills. This, if nothing else, prompts the question of whether laboratory work is being burdened with responsibilities it cannot realistically hope to meet.

Secondly, qualities such as open-mindedness, objectivity and a willingness to suspend judgement can hardly be claimed as uniquely scientific. Any scholar, acting in a scholarly capacity, is required to display these qualities, at least some of the time, and in science, as in other fields of creative activity, passionate personal conviction in the face of contrary evidence, rather than cold objectivity is sometimes needed to carry the day. Indeed, I often wonder whether this notion of the scientist as someone who, in Dainton's words, is 'objective to an objectionable degree' is not one that the scientific community would do well to discourage (Dainton 1971). In everyday life, of course, 'a willingness to suspend judgement' is not necessarily an asset.

Thirdly, as I have already indicated, there are formidable obstacles in providing students with an insight into scientific method and perhaps even more so at school level in developing expertise in using it. As the British Association for the Advancement of Science recognised long ago, 'scientific method', however it may be described, is appropriate only when addressing scientific problems (BAAS 1917). Most of the problems we meet in everyday life are not scientific and cannot be dealt with by scientific means. Within practical science at school, most of the problems which students meet are at best pseudo scientific, engaging the attention of some but certainly not all. Except when students are involved in genuine scientific investigations, the problems they encounter, or more commonly are presented with, are closed problems, with students often knowing the answer before the task is undertaken and the work becoming a lengthy elaboration of the obvious.

Fourthly, I think it is important to ask whether the manipulative skills that practical work in science often seeks to promote are skills that anyone is likely to need. While there perhaps was a time when the school science laboratory was a recognisable near relation of the research laboratory, that time is no longer. Is practical work in schools stuck with equipment, and I do not simply mean Bunsen burners and tripods, that is now unique to school science? Are students being taught to use techniques that, in the world beyond school, have long since been automated or rendered redundant in some other way?

Finally, as a comment on the broad categories of aims, we should ask whether some of them might not be better achieved in other ways. If we are serious about interesting young people in science in this electronic age, is the laboratory the obvious place to try and do it? Is it even the first port of call?

The ultimate question to be asked about aims, of course, especially those of such long standing as those associated with laboratory work, is whether they are being achieved. To answer this, it is necessary to turn to the research evidence. The picture, perhaps inevitably, is confused, reflecting not only different types of practical work, but a variety of methodologies, conflicting results from replicated studies, a degree of gender differentiation and work that ranges over motivation, skills acquisition, attitudes and concept development. There are some interesting omissions, for example, about the economics of practical work and the influence of factors as diverse as teachers' view of science and architectural space on the kinds of activities undertaken. Hodson, reviewing the research literature in 1992, came to the conclusion that 'It seems that all that can be safely concluded ...is that *some* teachers are able to use practical work successfully, with *some* students, to achieve *some* of their goals' (Hodson 1992: 96-7). White, writing four years later and exploring 'the link between the laboratory and learning' came to the rather pessimistic conclusion that 'There is insufficient evidence that laboratories promote better understanding of the methods of science and of abstraction and processes, make information memorable, reveal links between topics, and motivate, for them to withstand a determined attack should cost-cutting administrator decide to mount one' (White 1996 768). Even more recently, a study of the scientific investigations undertaken by students in England and Wales came to the conclusion that pupils were often 'unaware of the educational aims of investigational lessons' and that there was a 'mismatch between teachers' aims for investigation (over 50% of which were to teach procedures) and the things pupils consider they learnt during an investigation (only 20% of which referred to procedures)' (Watson, Goldsworthy and Wood-Robinson 1998: 20).

Redirecting practical work

I have been, like many others before me, more than a little critical of the claims made for the practical teaching of science. That was explicit in my title. I want to end on a more positive note by indicating the sorts of things that I think practical science should be about and which only practical science can achieve.

I would argue that it is only work at the bench or in the field that can give students what has sometimes been referred to as a feel for phenomena, a building up of experience about the natural phenomena that science seeks to understand and explain. Allied with this, I would also argue for at least some practical activities that help students understand how difficult it is to obtain secure knowledge about the natural world. At present, it seems to me that there is a risk that laboratory work may leave students with the view that finding out about the natural world is simply a matter of doing an experiment. It is not. It is about devising and planning that experiment and, therefore, about imagination, creativity, technique, persistence, collaboration, disappointment, difficulty, success and failure. It is about asking questions about what to measure, solving problems about how to measure it and evaluating the outcomes. To try and illustrate what I am getting at here, I will refer to three examples of practical work. The first is the familiar, if somewhat suspect, demonstration of a burning candle floating on water in a bell jar. As the candle burns and is eventually extinguished, the water slowly rises in the bell jar, until it the volume of air in the bell jar is only about 80 per cent of the original. From the teachers' point of view, it is a good demonstration. It always 'works' and offers a reliable demonstration that some twenty per cent of the air is oxygen. For the uninitiated pupil, it can, however, offer much more. When nine year old children were asked to describe what was going on and to try and explain what they saw, their explanations were very diverse and rich sources of speculation. The air became too damp for the candle to burn. The air in the bell jar became too smoky for the candle to keep burning. As the candle went out, the air inside 'shrunk'. The smoke from the candle 'went into' the water:- and so on. Each of these is, *a priori*, an interesting explanation and entirely consistent with the phenomena which the young children observed. How, in a science lesson, are children to be helped to choose between them and how often are they given a choice?

My second example refers to a 15 year old girl who, in her own words, 'liked biology' but 'hated physics and everything to do with it'. As part of her work in science, she chose to undertake a project in which she wanted to find out whether the rate of flow of water in a small river had any effect on the distribution of larvae attached to the underside of stones in the river. Her view, initially, was that the larvae would be swept away by faster moving water so that she expected a higher density of larvae where the water flowed more slowly. Her project, however, soon required her to measure the rate of water flow in different parts of a river, and to do so reliably and over time, since the level of the river varied with precipitation higher up the valley. I quote from her evaluation of her project in which she was required to record what she had learnt from her study.

> I learnt that science isn't just about doing experiments but you have to think and work out how to do things, so I learnt a lot about myself, like you have to stick at it and sometimes can't do what you want, so you do the next best thing. Measuring things isn't easy and I didn't always know what I was measuring. I'd like to go back and do it properly.

I think the conclusion I draw from this girl's work is that her project engaged her attention and taught her something not only about science but about herself and those seem to me laudable objectives.

My third, and final, example, reinforces this conclusion and I can deal with it briefly. The example is taken from the international Environment and School Initiatives Project and described by Peter Posch of the University of Klagenfurt (Posch 1993). In Italy, five upper secondary schools co-operated to study the quality and degree of pollution of ground and surface waters in the surrounding communities. The activities were co-ordinated by a group of students and teachers and financed by the communities on a contractual basis. The responsibilities of the students ranged from the selection and drawing of water samples, on the spot analyses, via a detailed chemical, bacteriological and microplankton analysis in their school laboratories to reporting and discussing the results with the authorities. The outcome of work of this kind is knowledge, which is not a reconstruction of existing knowledge but knowledge that is new in the sense that it provides information that was not hitherto available. As far as the students are concerned, the criteria for success seem clear. They must be involved in the evolution of their project work in science, and be given a chance to translate their values into action and to develop a sense of ownership of what they do.

My argument, therefore, is for involving all pupils in some kind of scientific investigation which meet these criteria at some stage during compulsory schooling. I do not underestimate the practical and other problems of achieving this but if the gains are thought great enough, these problems will be overcome. The gains are to do with pupil motivation, some insights into the nature of science, the benefits of collaborative working (perhaps involving the Internet), the deployment of different competences within a group, the acquisition of communication skills, the melting of disciplinary boundaries and the breaking down of the barriers between school and what is misleadingly called the real world. The alternative is to try and go on as we have been doing for over a century. The only issue is whether our students will continue to allow us to do so.

References

British Association for the Advancement of Science (1917). *Report*. London: Murray.
Dainton, F.S. (1971). *Science: Salvation or Damnation*. Southampton: University of Southampton.
Gibbons, M., Limoges, C., Nowotny, H., Schwartzman, S., Scott, P. & Trow, M. (1994). *The new production of knowledge. The dynamics of science and research in contemporary societies*. London: Sage
Gott, R. & Duggan, S. (1995). *Investigative Work in the Science Curriculum*. Buckingham: Open University Press
Hodson, D. (1993). Rethinking Old Ways; Towards a More Critical Approach to Practical work in School Science. *Studies in Science Education* **22**, 85-142
Irwin, A. & Wynne, B. (eds.) (1996). *Misunderstanding science? The public reconstruction of science and technology*. Cambridge: Cambridge University Press
Jasanoff, S., Markle, G.E., Petersen, J.C. & Pinch, T. (1995). *Handbook of Science and Technology Studies*. London: Sage
Laudan, L., Donovan, A., Laudan, R., Barker, P., Brown, H., Leplin, J., Thagard, P. & Wystra, S. (1986). Scientific change: philosophical models and historical research. *Synthese*, **69** (1), 141-223

Layton, D., Jenkins, E.W., Macgill, S. & Davey, A. (1993). *Inarticulate Science? Perspectives on the public understanding of science and their implications for the learning of science.* Driffield: Studies in Education.

Medawar, P. (1982). *Pluto's Republic.* Oxford: Oxford University Press

Phillips, D.C. (1995). The Good, the Bad and the Ugly; The Many Faces of Constructivism, *Educational Researcher,* 24, (7) 5-12.

Posch, P. (1993). Research Issues in Environmental Education. *Studies in Science Education* **21**, 21-48.

Ravetz, J.R. (1970). *Scientific Knowledge and its Social Problems.* Oxford; Clarendon Press.

Ravetz, J.R. (1990). New ideas about science, relevant to education. In E.W. Jenkins (ed.), *Policy Issues and School science Education.* Leeds: University of Leeds Centre for Studies in Science and Mathematics Education, pp.18-27.

Redner, H. (1987). *The Ends of Science: An Essay in Scientific Authority.* Boulder: Westview Press.

Roth, W-M. (1998). *Designing Communities.* Dordrecht: Kluwer

Solomon, J. (1980): *Teaching Children in the Laboratory.* London: Croom Helm

Watson, R., Goldsworthy, A. & Wood-Robinson, V. (1998). Getting AKSIS to Investigations. *Education in Science,* **177**, 20-21.

White, R.T. (1996). The link between the laboratory and learning. *International Journal of Science Education,* **18**, 7, 761-774

'Mapping' the domain
Varieties of practical work

Robin Millar, Jean-François Le Maréchal and Andrée Tiberghien

Practical work occupies a central place in science education in many countries. Practical work in science, however, is very varied in type, and in intention. If we, as teachers and researchers, want to explore the effectiveness of practical work in achieving educational goals, then we need to be clear about the different types of practical work which are (or could be) undertaken in classes, and their different purposes and characteristics.

In this paper, a typology of practical work (a 'map') is presented, and some of its implications for teaching and research will be explored. A 'map' of this sort may help us see how to address the key question of the effectiveness of practical work.

1 Introduction

The subject matter of science is the natural world around us, what it contains, how it works, and how we can explain and perhaps predict its behaviour. So it is hardly surprising that, in teaching science, we often find that we want to allow students to observe, handle and manipulate objects and materials for themselves, rather than simply talking about them, or showing representations of them (such as diagrams, photographs, or video extracts). Not only does practical work with real objects and materials help us to communicate information and ideas about the natural world, it also provides opportunities to develop students' understanding of the scientific approach to enquiry. Furthermore, practical work is the aspect of science education which students say they most enjoy. For all these reasons, most science educators would agree that science courses at school and university levels should contain significant amounts of practical work.

Practical work carried out by the students themselves, usually working in small groups, is a prominent feature of school science education in many countries. In other countries where there has not been such a strong tradition of student practical work, introducing it, or increasing the amount of it, is often seen as a desirable reform, and as a means of improving science education. This enthusiasm for practical work in science education persists despite the fact that it makes science a relatively expensive subject on the school curriculum. Laboratories are more expensive to build and maintain than ordinary classrooms; practical work requires equipment, and materials which are used up in the teaching activities and have to be replaced.

In countries where there is an established tradition of student practical work in school science, however, its effectiveness is being increasingly questioned. Students often fail to learn from practical work the things we had intended them to learn. Frequently it is carried out very rapidly, or with unreliable equipment, or with insufficient attention to care and precision, so that students fail even to produce the phenomenon they are supposed to observe, let alone be helped to appreciate patterns, trends or explanations. Even when the outcomes are as the teacher intended, conclusions which seem 'obvious' to the teacher can appear less so to the student. Often the work is humdrum and routine, rather than engaging or inspiring. In one recent critique of laboratory practical work, Hodson (1991) argues that:

> As practised in many schools, it [practical work] is ill-conceived, confused and unproductive. For many children, what goes on in the laboratory contributes little to their learning of science or to their learning about science and its methods. Nor does it engage them in doing science in any meaningful sense. At the root of the problem is the unthinking use of laboratory work. (p. 176)

The issue, then, is not about the usefulness, or otherwise, of practical work in general. For the reasons outlined in the opening paragraph, practical work is likely to remain a prominent part of science education. Instead we need to ask about the effectiveness of *specific* pieces of practical work for achieving *specific* learning outcomes. The aim of the work presented in this article was to develop a framework which can help us to raise such questions and, in time perhaps, to answer some of them.

This framework consists of a 'map' – or classification system – which allows us to describe in detail the characteristics of any given piece of practical work. The 'map' of practical work described in this article is a slightly modified version of a 'map of labwork' which was developed by the authors as a general framework for use in the European project *Improving Labwork in Science Education* (LSE) (Millar, Le Marechal and Tiberghien, 1998).

Although a number of classification schemes have been published in the science education literature for analysing types of interaction in the science teaching laboratory (Eggleston, Galton and Jones, 1975; Shymansky et al., 1976; Penick et al., 1976; Hacker, 1984; Hegarty-Hazel, 1990; Giddings, Hofstein and Lunetta, 1991), there have been very few attempts to develop a classification system for science laboratory tasks themselves. Some categories, of course, are well established, such as teacher demonstration and student practical work. The idea of 'levels of enquiry', to describe whether the problem, the procedure and the conclusion are open or given (Herron, 1971), has become widely used. Woolnough and Allsop (1985) propose a general classification of practical tasks into four groups: exercises, experiences, investigations, and illustrations of theory. Kirschner and Meester (1988) suggest a four way classification of laboratory approaches: formal (to illustrate laws and concepts), experimental (open-ended), divergent (from a common start), and skills/procedures related. More detailed classification schemes have also been proposed. Lunetta and Tamir (1981) used a modified version of an earlier scheme by Fuhrman et al. (1978) to compare laboratory tasks in the *PSSC Physics* course and the *Project Physics Course*. More recently McComas (1997) has proposed a taxonomy of 'physical factors' (aspects of the laboratory, the curriculum) and 'personal factors' (characteristics of students and teachers).

Any classification system is designed for a particular purpose and builds in, to some extent, the perspectives of its constructors. In our view, the core purpose of practical activity in science teaching is to help the student make links between the domain of objects and observable things, and the domain of ideas. So we wanted to develop a classification system for practical tasks which would focus more prominently on the kinds of physical actions and operations required of the student in dealing with objects and observables on the one hand, and the kinds of mental actions and operations required of the student in dealing with ideas on the other. The overall purpose is to develop a tool which can be used to provide the kind of information which may be useful for thinking about how to modify and improve practical tasks.

In the LSE project, labwork was defined in a way, which included some activities that might not normally be regarded as practical work, and the focus was on teaching at upper secondary school and university level. In this chapter we have modified some categories and added some new ones to make the map better suited to classifying the kinds of practical tasks used at primary and secondary school level.

Robin Millar, Jean-François Le Maréchal and Andrée Tiberghien

2   The domain of practical work

It may be useful to begin by making clearer exactly what we will include within the category of 'practical work'. We might define practical work as:

> all those teaching and learning activities in science which involve students at some point in handling or observing the objects or materials they are studying.

This places no restrictions on where the work is carried out. Practical work might be carried out in a laboratory, or outside 'in the field', or in an ordinary classroom. By including the words 'at some point' in the definition above, we emphasise that practical work involves conceptual activity as well as practical activity, so that observing or handling objects and materials is just one element of a practical task. The definition above would also include teaching and learning activities in which the students watch someone else (often the teacher) handle objects or materials, as well as those in which they handle them for themselves, i.e. it includes teacher demonstrations as well as pupil practical work.

We might also want to extend this definition to include activities in which students worked with representations of real objects or materials, such computer simulations, or video recordings of events which would be too dangerous, or too expensive, or too difficult to work with 'for real'. Our definition of practical work would then be enlarged to:

> all those kinds of learning activities in science which involve students at some point in handling or observing real objects or materials (or direct representations of these, in a simulation or video-recording) .

There are, however, many ways of classifying any large collection of items – and the merits of any particular classification system depend on the purpose for which it was developed. Our aim, as we have said above, is to provide a framework for asking more precise and specific questions about the effectiveness of practical work. So we will now go on to look a little more closely at what we might mean by 'effectiveness' in this context.

'Mapping' the domain

3 The effectiveness of a practical task

To help us clarify what we mean by the 'effectiveness' of a practical task in a teaching and learning situation, it is useful to consider the processes involved in designing and evaluating a practical task. A possible model of the logical steps in this process, and some of the influences on them, is shown in Figure 1.

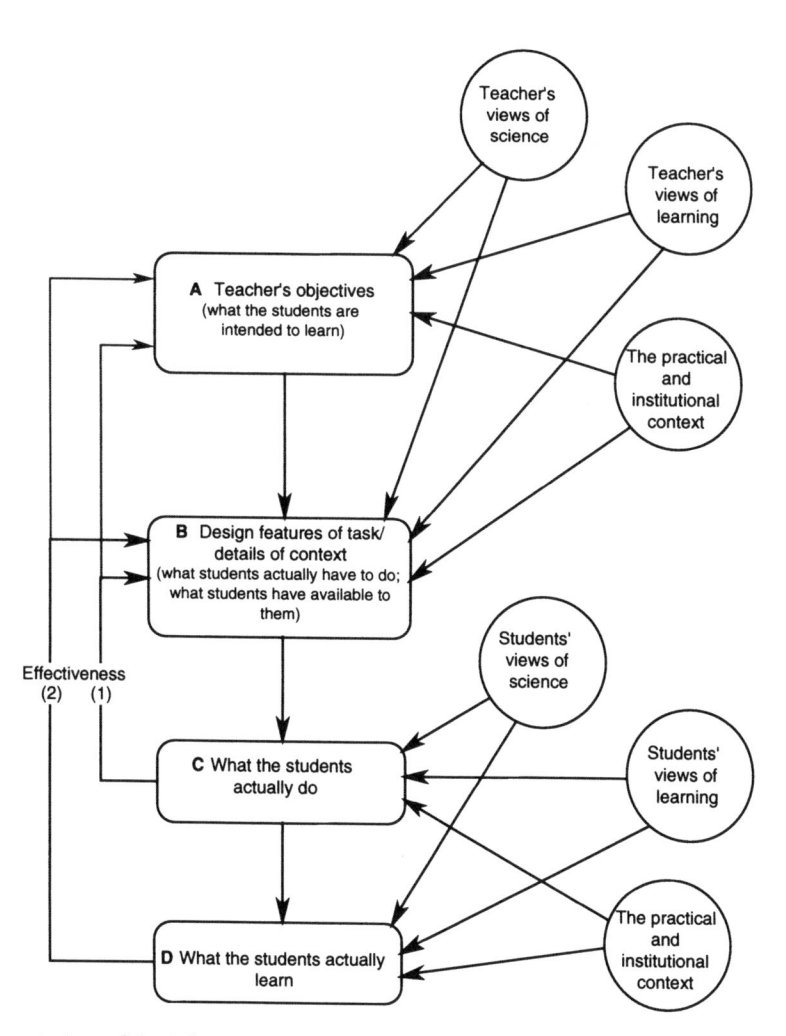

Figure 1 *A model of the process of design and evaluation of a teaching and learning task*

Robin Millar, Jean-François Le Maréchal and Andrée Tiberghien

The starting point is Box A: the learning objectives the teacher has in mind. What does he or she want the students to learn? This then leads on to the design of the practical task which is to be used to achieve those objectives (Box B). The choice of learning objectives and the decisions taken about the design of the task (Boxes A and B) are influenced by many considerations. Three of these are shown: the teacher's view of science (ideas about what is important to try to teach, about the nature of this knowledge, and so on); the teacher's view of learning (ideas about how students learn); and the practical and institutional context (the facilities, resources and time available, the way in which students will be assessed, and so on). Some aspects of this background influence may be explicitly acknowledged in selecting the objectives and designing the task, whilst others may be tacit influences.

In designing a practical task, the teacher intends that the students will *do* something when given the task. So the model in Figure 1 leads on to the question of what the students *actually* do when carrying out the task (Box C). The students may do what the teacher intended, or something which differs from it to a greater or lesser extent. For example, students may misunderstand the instructions and carry out actions which are not the ones the teacher intended. Or they may carry out the intended operations on objects and materials, but not engage in the kind of thinking about these which the teacher intended. Finally, we move on to box D: what do the students actually learn from carrying out the task? Like the teacher's decisions in planning the task, the students' actions and learning as they carry it out (Boxes C and D) are also influenced by many factors, three of which are their views of science (their interest in the subject matter, their understanding of the connections to other ideas, and so on), their views of learning (ideas about how one learns the sorts of ideas involved), and the practical and institutional context of the task (the quality of the equipment, the time available, the importance of the task in relation to achieving success in the course, and so on).

This model (Figure 1) now allows us to say more precisely what we mean when we ask: how effective is a particular practical task? We can distinguish two separate questions. First, do the students actually do the things we wished them to do when we designed the task? This is about the relationship between Boxes C and B. We refer to this as 'Effectiveness (1)'. This then leads on to the question of the effectiveness of a task in promoting student learning (the relationship between D and A). This we call 'Effectiveness (2)'.

If we collect some evidence which suggests that a practical task is not

'Mapping' the domain

very effective, in terms of either Effectiveness (1) or Effectiveness (2), then we might want to re-design the task, whilst keeping the learning objectives the same – or we might feel that we need also to reconsider our objectives, perhaps to make these clearer or less ambitious.

In order to do this, however, we need to have a clear idea about two things:

- the intended learning outcomes, or learning objectives, of the task
- the range of possible variation in the design of the task and the context within which it is carried out.

This allows us to characterise the task we are considering, and may suggest ways in which its objectives or design could be altered. The 'map' which we will describe in the next section provides a means of doing this.

4 A 'map' of practical work

The model set out in Figure 1 identifies the two major dimensions of a 'map', or classification system, for practical tasks. The first is the intended learning outcome, or learning objective, of the task. Here we want to classify tasks according to their main learning objective(s). The second dimension is the task design itself. This is more complex and can be sub-divided into a number of sub-dimensions. Any given task can be assigned to one (or more) coding categories within each of these sub-dimensions. The structure of our 'map' is shown in outline in Figure 2.

Many of these dimensions and sub-dimensions are self-explanatory. B1.1 to B1.3, however, may need some explanation. The idea which underlies these is that the fundamental purpose of practical work is to help students to *build bridges between two distinct domains*: the domain of real objects and observable things, and the domain of ideas (Figure 3).

So, in any practical task, students are expected to do certain kinds of things with objects and observables; and they are also expected to do certain things with ideas. Sub-dimensions B1.1 and B1.2 refer to these two aspects of the practical task, whilst sub-dimension B1.3 asks: does the work on objects lead towards the work on ideas (in an inductive manner), or do the ideas come first and lead to the work on objects (perhaps in a more hypothetico-deductive manner)?

A	*Intended learning outcome (learning objective)*
B1	*Design features of the task*
	B1.1 What students are intended to do with objects and observables
	B1.2 What students are intended to do with ideas
	B1.3 Whether the task is objects- or ideas-driven
	B1.4 The degree of openness/closure of the task
	B1.5 The nature of student involvement in the task
B2	*Practical context of the task*
	B2.1 The duration of the task
	B2.2 The people with whom the student interacts whilst carrying out the task
	B2.3 Information given to the student on the task
	B2.4 The type of apparatus involved
B3	*Student's record of work on the task*
	B3.1 Nature of record
	B3.2 Purpose of record
	B3.3 Audience for record

Figure 2 *Main dimensions and sub-dimensions of a 'map' of practical work tasks*

domain of real objects and observable things ↔ domain of ideas

Figure 3 *Labwork: helping students to make links between two domains*

For each sub-dimension, we can then characterise a practical task either by selecting the most appropriate descriptor (or descriptors) from a list, or by ticking the appropriate boxes in a table. The coding categories within each sub-dimension are not mutually exclusive; many practical tasks will have elements which match several of the coding categories. Our aim is not to provide a set of 'pigeon-holes' for each sub-category, so that every task can fit neatly into one of them. Rather it is to provide a means of obtaining a characteristic 'profile' of each practical task.

'Mapping' the domain

5 Coding categories for each sub-dimension

We will now discuss the coding categories we propose within each sub-dimension, providing examples in some sub-dimensions where this will help to clarify the meaning of the coding categories.

5.1 Dimension A: Intended learning outcome (learning objective)

A first dimension to consider in classifying a practical task is the intended learning outcome (or learning objective) which the teacher or the designer of the task has in mind in presenting the task. Learning objectives fall into two main categories, associated with the learning of science content or of the processes of scientific enquiry. These can be further sub-divided as shown in Table 1.

Content:

a	to help students identify objects and phenomena and become familiar with them	
b	to help students learn a fact (or facts)	
c	to help students learn a concept	
d	to help students learn a relationship	
e	to help students learn a theory/model	

Process:

f	to help students learn how to use a standard laboratory instrument, or to set up and use a standard piece of apparatus	
g	to help students learn how to carry out a standard procedure	
h	to help students learn how to plan an investigation to address a specific question or problem	
i	to help students learn how to process data	
j	to help students learn how to use data to support a conclusion	
k	to help students learn how to communicate the results of their work	

Table 1. *Dimension A: Intended learning outcome. Coding categories*

Some of the terms used here may be worth clarifying briefly. In b, a 'fact' means a statement which can be readily agreed, such as that pure water

41

boils at (or near to) 100°C, or that common salt dissolves in water whilst chalk does not. In d, a 'relationship' might be a pattern or regularity in the behaviour of a set of objects or substances, or an empirical law.

Many practical tasks are likely to have more than one of these learning objectives. It is also unlikely that some (like k) would ever be the sole objective of a practical task. In classifying a task by its learning objective(s), it is more useful if the focus is on the most important objective or objectives rather than identifying all the possible objectives which the task might be said to address.

5.2 Dimension B1: Design features of task

The second dimension along which practical tasks need to be classified concerns the design features of the task.

Sub-dimension B1.1: What students are intended to do with objects and observables

Some practical tasks require students to *use* an instrument, or a laboratory device, or a standard laboratory procedure. Others ask students to *present* an object so as to *display* certain features of it clearly, for example in a dissection of a flowering plant. Some practical tasks require the student to *make* something, for example a physical object or a material (e.g. a chemical substance), or to make an event occur. The fourth, and perhaps the largest, category of practical tasks is those which require the student to *observe* something. The observation may be of an object, or of a material, or of an event, or of a physical quantity (or variable) associated with an object, or material, or event. Also this observation may be qualitative (e.g. an observation of colour), or semi-quantitative (noting if something is large, or small), or quantitative (i.e. a *measurement*). Some examples of practical tasks of each of these kinds are shown in Appendix 1.

This aspect of a practical task can be coded using Table 2. Note how this table can also be used to provide additional information, by indicating whether the source of data which the students acquire comes from the real world, inside or outside the laboratory, or from a representation of the real world, such as a video recording or computer simulation. Also, in the case of the 'observe' categories, we can indicate whether the observation is qualitative, semi-quantitative, or quantitative, by using the codes Ql, SQt, and Qt, rather than a simple tick, in the appropriate boxes.

'Mapping' the domain

What students are intended to do with objects and observables		Source of data acquired by students (tick as appropriate)			
		from real world		from video	from computer or CD-ROM
		inside laboratory	outside laboratory		
use	an observation or measuring instrument				
	a laboratory device or arrangement				
	a laboratory procedure				
present or display	an object				
make	an object				
	a material				
	an event occur				
observe	an object				
	a material				
	an event				
	a quantity				

Table 2 *Sub-dimension B1.1: What students are intended to do with objects and observables. Coding categories*

In Table 2, although all the coding categories are expressed in the singular (an object, a material, etc.), this should also be taken to include the plural, for example, where a practical task involves making several objects or materials, or observing more than one object, material, event, or quantity. Also, for many practical tasks, it is obvious that more than one box in Table 2 will apply. For example, measuring a physical quantity ('observe a quantity') necessarily involves using a measuring instrument. But for some tasks coded as 'observe a quantity' this will not be the case. So the combination of codes gives a fuller picture of the requirements of the task.

Robin Millar, Jean-François Le Maréchal and Andrée Tiberghien

Sub-dimension B1.2: What students are intended to do with ideas
Practical tasks do not only involve observation and/or manipulation of objects and materials. They also involve the students in using, applying, and perhaps extending, their ideas. That is, in addition to 'work with the hands', practical work also requires 'work with the head'. As we have said above, the central role of practical work in science education stems from its power to bridge the two domains of observables and ideas. So a practical work task can also be classified according to what the students are intended to do with ideas.

Some practical tasks simply require *direct reporting* of observations, though, of course, the selection of features to observe and record is inevitably influenced by the teacher's and/or the student's ideas about the task and its purpose. Other tasks require the student to *identify a pattern*, or regularity, in the behaviour of the objects or events observed. One particular type of 'pattern' which is common (and so worth keeping as a separate category) is a *relationship between objects*, or between *physical quantities* (variables). Another type of practical task requires students to *'invent' or 'discover' a new concept*. We find it difficult to give a concise label for this category, and might seem to be hedging our bets between an empiricist view and a radical constructivist one by using both the words 'invent' and 'discover'. We are thinking here of practical tasks in which the first step is to realise the need for a new parameter which allows a model (usually a quantitative one) to fit better with the data; the second step is then to construct a meaning for this parameter which then becomes a 'new' physical quantity, or 'new' concept. We think this second stage is likely to be teacher-mediated in almost all cases. Practical work of this type is very rare at school level, though it may be more common in the research context.

Some practical tasks focus on *determining the value of a quantity*, using an indirect method. This is distinctly different from direct measurement using a single measuring device. Here students are applying their ideas to obtain a numerical value of the quantity from a number of other, more fundamental quantities, which *can* be measured directly. Another group of tasks is those which involve *testing predictions*. In such tasks, a prediction may be simply a guess, or it may be deduced from a more formal understanding of the situation, such as an empirical law, or a theory (or model). We are using the word 'testing' loosely here, to mean seeking a match between prediction and observation. We do not want to imply that we think practical tasks in the teaching laboratory provide 'severe tests' of well-established ideas! Usually the real task for the student is to 'produce the phenomenon', that is, to

'Mapping' the domain

succeed in producing the outcome predicted by a well-established scientific explanation. Finally, some practical tasks are about *accounting for observations*, either by relating them to a given explanation or by proposing an explanation. An 'explanation' might be an empirical law, or a general theory, or a model derived from a general theory, or general principles derived from a theoretical framework. In some tasks, the explanatory ideas are known in advance and the student is expected to use these to account for what is observed, perhaps extending or modifying the framework of ideas. A variant of this is where two (or more) possible explanations are proposed and the task is to decide which accounts better (or best) for the data. In other tasks, the observations come first, and the student is expected to select an explanation from his/her existing knowledge, or perhaps to extend this to develop an ex-

What students are intended to do with ideas		Tools to be used for processing information (tick as appropriate)		
		manual	pocket calculator	computer
report observation(s)				
identify a pattern				
explore relation between	objects			
	physical quantities (variables)			
	objects and physical quantities			
'invent' (or 'discover') a new concept (physical quantity, or entity)				
determine the value of a quantity which is not measured directly				
test a prediction	from a guess			
	from a law			
	from a theory (or a model based on a theoretical framework)			
account for observations	in terms of a given explanation			
	by choosing between two (or more) given explanations			
	by proposing an explanation			

Table 3   *Sub-dimension B1.2: what students are intended to do with ideas. Coding categories*

a	What the students are intended to do with ideas arises from what they are intended to do with objects;	
b	What the students are intended to do with objects arises from what they are intended to do with ideas	
c	There is no clear relationship between what the students are intended to do with objects and with ideas	

Table 4 *Sub-dimension B1.3: objects- or ideas- driven? Coding categories*

Aspect of practical task	Specified by teacher	Decided through teacher-student discussion	Chosen by students
	(tick as appropriate)		
Question to be addressed			
Equipment to be used			
Procedure to be followed			
Methods of handling data collected			
Interpretation of results			

Table 5 *Sub-dimension B1.4: degree of openness/closure of a practical task. Coding categories*

planation. Some examples of practical tasks of each of these kinds are shown in Appendix 2.

This aspect of a practical task can be coded using Table 3. Again this table can be used to provide additional information, by placing ticks in the appropriate column to indicate the kinds of tools available to students for processing the information they obtain.

Sub-dimension B1.3: Objects- or ideas-driven?
A third sub-dimension of the design of a practical task concerns the relationship between the two domains: of objects and observables, and of ideas. Some tasks are presented in an 'objects-driven' way: the student is required to carry out some operations on objects from which, it is hoped, ideas will emerge. Other tasks are presented in an 'ideas-driven' way: the operations on objects are specifically undertaken to explore some ideas which have been stated in advance. Of course, to some extent, all observation is guided by the

'Mapping' the domain

ideas of the observer (or the teacher giving the instructions). This dimension of the 'map' is intended simply to reflect the *emphasis* in the practical work task. The coding categories are shown in Table 4.

Sub-dimension B1.4: Degree of openness/closure
Practical tasks can differ widely in the extent to which the student is able (or required) to take decisions about aspects of the task. So the fourth sub-dimension of the design of a practical task distinguishes between 'open' and more 'closed' tasks, by looking at who takes the decisions about different aspects of the design and conduct of the task. The pattern of ticks in Table 5 provides an indication of the degree of openness or closure of the task.

Sub-dimension B1.5: Nature of student involvement
The fifth and final sub-dimension of task design concerns the nature of student involvement in the practical task. The coding categories are shown in Table 6.

a	demonstrated by teacher; students observe	
b	demonstrated by teacher; students observe and assist as directed (e.g. in making observations or measurements)	
c	carried out by students in small groups	
d	carried out by individual students	

Table 6 *Sub-dimension B1.5: nature of student involvement. Coding categories*

5.3 Dimension B2: Context of the task

In addition to the wide variations possible in the design of practical tasks, there can also be wide differences in the context within which the task is carried out. Four sub-dimensions relating to task context can be identified: the amount of time that is given to the task, the people with whom the student is expected to interact whilst carrying out the task, the way in which information is given to the student on the task, and the type of apparatus involved in carrying out the task.

Compared with dimension B1 of the 'map', these dimensions and their associated coding categories are largely self-explanatory. They are summarised in Tables 7-10.

a very short (less than 20 minutes)	
b short (one science lesson, say, up to 80 minutes)	
c medium (2-3 science lessons)	
d long (2 weeks or more)	

Table 7 *Sub-dimension B2.1: duration of task. Coding categories*

a other students carrying out the same practical task	
b other students who have already completed the task	
c teacher	
d more advanced students (demonstrators, etc.)	
e others (technician, glassblower, etc.)	

Table 8 *Sub-dimension B2.2: people with whom the student interacts. Coding categories*

a oral instructions	
b instructions on blackboard/whiteboard/OHP	
c guiding worksheet	
d textbook(s)	
e other (e.g. data book, data base, instruction manual, etc.)	

Table 9 *Sub-dimension B2.3: information given to the student on the task. Coding categories*

a demonstrated by teacher; students observe	
a standard laboratory equipment	
b standard laboratory equipment + interface to computer	
c everyday equipment (kitchen scales, domestic materials...)	

Table 10 *Sub-dimension B2.4: type of apparatus involved. Coding categories*

5.4 Dimension B3: The student's record of work on the task

Finally, practical tasks can differ in the ways in which the student is expected to produce a record of work on the task, the purposes of such a record and the audience for whom it is produced. As with dimension B2, the sub-dimensions of B3 and their associated coding categories are largely self-explanatory. They are summarised in Tables 11-13.

a	no written record	
b	notes	
c	completion of printed worksheet	
d	written account (using given structure and format)	

Table 11 *Sub-dimension B3.1: nature of student's record of work on the task. Coding categories*

a	to assist students in learning science content or process	
b	to provide evidence that the task has been carried out	
c	as a basis for assessing the student's performance	
d	as a record which the student can use to revise for tests or examinations	
e	to help students learn how to write a scientific report	

Table 12 *Sub-dimension B3.2: purpose of student's record of work on the task. Coding categories*

a	the student	
b	teacher	
c	other students	
d	other	

Table 13 *Sub-dimension B3.3: audience for student's record of work on the task. Coding categories*

6 Possible uses of a 'map' of practical work

A coding proforma for describing in detail the characteristics of any practical task can be assembled by collating Tables 1-13 above. This can be used to produce a profile of the salient features of any given practical task. What, then, might be this classification and coding system be used for? First, it can be used to identify similarities and differences in the kinds of practical work used in school science courses, perhaps to compare the types used at different ages or stages, or in the different science disciplines. In this way, for example, we might discover that some types appeared to be underused, or overused. The coding categories may also suggest questions which we might ask about practical work. Does a course make as much use as we might wish of more open-ended tasks? What is the degree of openness in the tasks used, and how does this compare with our intentions?

The 'map' also provides us with a clearer basis for addressing questions of the effectiveness of practical work. From this discussion of dimensions, sub-dimensions and coding categories, it should be clear that there is a very wide variety of types of practical task which might be used in teaching and learning science. It clearly does not make sense to ask in general: is practical work an effective means of teaching science? For any given task, however, we can use the 'map' to provide a full description of the characteristics of the task. Then we are in a better position to carry out research to see whether students actually do the things (with objects and with ideas) that we want them to – and, if we find they do not, we can consider whether adjustments to the task design or the context within which it is carried out, by modifying it on one or more of the sub-dimensions which relate to these, might make it more effective in attaining its goals. In this way we may be able, over time, to make gradual progress towards a more efective use of practical work in science courses at all levels.

Acknowledgements

The 'map' of practical work described in this article is a slightly modified version of the 'map of practical work' which was developed by the authors as part of the work of the project *Improving Labwork in Science Education* (LSE) (Project PL95 2005), supported by the European Commission DGXII. We are grateful for this support, and also acknowledge the contribution of our colleagues in the LSE project to our ideas as we developed the 'map'.

References

Eggleston, J.F., Galton, M.J. and Jones, M. (1975). *A Science Teaching Observation Schedule*. Schools Council Research Series. London: Macmillan.

Fuhrman, M., Novick, S., Lunetta, V. and Tamir, P. (1978). *A laboratory organisation and task analysis inventory.* (Technical Report 15). Iowa City, IA: Science Education Center, University of Iowa.

Giddings, G.J., Hofstein, A. and Lunetta, V. (1991). Assessment and evaluation in the science laboratory. In B. Woolnough (Ed.) *Practical Science* (pp. 167-177). Buckingham: Open University Press.

Hacker, R.G. (1984). A typology of approaches to science teaching in schools. *European Journal of Science Education*, 6 (2), 153-167.

Hegarty-Hazel, E.H. (1990). Tertiary science classrooms. In E.H. Hegarty-Hazel (Ed.) *The Student Laboratory and the Science Curriculum* (pp. 357-382). London: Routledge.

Herron, M.D. (1971). The nature of scientific enquiry. *School Review*, 79, 171-212.

Hodson, D. (1991). Practical work in science: Time for a reappraisal. *Studies in Science Education*, 19, 175-184.

Kirschner, P. and Meester, M. (1988). Laboratory approaches. *Higher Education*, 17 (1), 81-98.

Lunetta, V.N. and Tamir, P. (1981). An analysis of laboratory activities: Project Physics and PSSC. *School Science and Mathematics*, 81, 635-642.

McComas, W.F. (1997). The laboratory environment: An ecological perspective. *Science Education International*, 8 (2), 12-16.

Millar, R., Le Maréchal, J-F. and Tiberghien, A. (1998). *A Map of the Variety of Labwork*. Labwork in Science Education (LSE) Project. Working Paper 1. European Commission TSER Programme Project PL 95-2005. ISBN 0-904-42191-0.

Penick, J.E., Shymansky, J.A., Filkins, K.M. and Kyle, W.C. (1976). *Science Laboratory Interaction Categories (SLIC) – Student*. Iowa City, IA: Science Education Center, University of Iowa.

Shymansky, J.A., Penick, J.E., Kelsey, L.J. and Foster, G.W. (1976). *Science Laboratory Interaction Categories (SLIC) – Teacher*. Iowa City, IA: Science Education Center, University of Iowa.

Woolnough, B. and Allsop, T. (1985). *Practical Work in Science*. Cambridge: Cambridge University Press.

Robin Millar, Jean-François Le Maréchal and Andrée Tiberghien

Appendix 1 Coding categories for sub-dimension B1.1 with examples of each

What students are expected to do with objects and observable things	Examples
use an observation or measuring instrument	*use a microscope to look at onion skin cells* *use a cathode ray oscilloscope (CRO) to look at some signal waveforms* *use a burette to deliver measured volumes of a liquid*
use a laboratory device or arrangement	*set up distillation apparatus to separate two miscible liquids* *use a dissecting kit/scalpel to remove a muscle from a chicken wing* *set up a filter funnel to separate a solid from a liquid*
use a laboratory procedure	*carry out a recrystallisation of a compound to produce a purer sample* *set up a control for a biological investigation* *follow a standard schedule for qualitative analysis of a sample of an unknown inorganic compound*
present or display an object	*carry out a dissection of a biological specimen to display the main features of interest* *display a collection of geological specimens to illustrate a particular feature*
make an object	*make a microscope slide to display the cells of a given specimen* *make an electric circuit from a given circuit diagram*
make a material	*synthesise a particular chemical substance*
make an event occur	*tune an electric circuit containing a capacitor (C) and an inductor (L) to show resonance*
observe an object	*note and record the pattern of iron filings sprinkled around a bar magnet* *look at some fossil specimens* *inspect some rock samples with a hand lens for evidence of volcanic origins*
observe a material	*note and record the shape of crystals of copper sulphate* *note and record the physical properties of a sample of polythene*

'Mapping' the domain

observe an event	record the manner in which an animal (an invertebrate, a fish) moves
	note what happens when a piece of sodium is placed in water
	pass a ray of white light through a prism and note the spectrum produced
	make observations of the germination and growth of a broad bean
	note whether an object floats or sinks when placed in water
observe a quantity	measure the resistance of a piece of wire
	measure the volume of an acid solution needed to neutralise a given volume of an alkali solution
	measure the density of a sample of a solid material
	measure the length of a spring with different loads hanging from it
	measure the melting point of a substance
	observe the change in temperature of water in an insulated container over a period of time

Robin Millar, Jean-François Le Maréchal and Andrée Tiberghien

Appendix 2   Coding categories for sub-dimension B1.2 with examples of each

What students are expected to do with ideas	*Examples*
report observation(s)	*describe in detail how a fish moves* *describe the shape of crystals of a given substance*
identify a pattern	*compare the measured IR spectrum of an organic compound to known IR spectra in order to identify it* *note the different plant and animal species found at different levels of a seashore habitat.* *compare the outcome of a test for glucose on a sample of foodstuff with a previously observed positive test* *note the regular changes in the appearance of the moon over a 29 day cycle* *identify the objects and variables in an environment which are involved in some interactions of interest within that environment*
explore relation between objects	*note that a pinhole camera produces an inverted image on the screen*
explore relation between physical quantities	*find out how the [extension – increase of length] of a spring depends on the [load -mass] attached to it* *find out how temperature and concentration of reagents affect the rate of an enzyme reaction*
explore relation between objects and physical quantities	*compare rates of reaction of a selection of metals with dilute acid* *investigate the effect of different drinks (tea, coffee, cocoa, etc.) on rate of heartbeat*
'invent' (or 'discover') a new concept (physical quantity, or entity)	*identify the need for (or the usefulness of) the quantity defined as energy/time (power) in accounting for a set of observations* *identify that, for a weak acid, the ratio: log ($[H^+].[A^-]/[HA]$) is constant and can be used to characterise the strength of the acid*
determine the value of a quantity which is not measured directly	*measure the acceleration due to gravity using the relationship $T = 2\pi R(\F(l,g))$ for a simple pendulum* *measure the thermal conductivity of a material* *determine the number of molecules of water of crystallisation associated with each molecule of a salt using volumetric analysis*
test a prediction based on a guess	*test the prediction that rubber-soled shoes provide better 'grip' on a wooden floor* *guess that a sample of soil is mainly limestone and test for effervescence when dilute acid is added*

test a prediction from a law	*test whether the increase in length (extension) of a piece of elastic is directly proportional to the load applied, up to a limit (as predicted by Hooke's Law)* *test whether the electric current through a given conductor is proportional to the applied p.d. (as predicted by Ohm's Law)*
test a prediction from a theory (or a model based on a theoretical framework)	*a model of the mechanism of a chemical reaction predicts a certain relationship between rate of reaction and temperature; test this by comparing it with what is actually observed* *predict the pH of a solution of ethanoic acid of given concentration using the formula: $pH = \frac{1}{2}(pK_a - \log[c])$ and then check this by measurement* *theory of fluid flow predicts that the volume of fluid per second (Q) flowing through a pipe is related to pressure difference (P), radius (r) and length (l) by the equation: $Q = \frac{P\pi r^4}{8\eta l}$, where h is the viscosity of the liquid. Carry out a series of measurements to test this relationship in the case of water*
account for observations in terms of a given explanation	*explain similarities and differences between related species of birds in terms of a given account of their evolution* *explain observations in some displacement reactions (metal/metal salt solution) in terms of the reactivity series for metals*
account for observations by choosing between two (or more) given explanations	*is the behaviour observed when the temperature of a sample of air is raised better explained by saying that 'hot air rises' or 'air expands when heated'*
account for observations by proposing an explanation	*from observation of the objects and variables in an environment, propose a model to explain some aspect of the interactions within that environment* *measure the temperature of a sample of water in a calorimeter over a period of minutes as it is heated by an immersion heater. Explain the shape of the temperature-time graph produced*

Robin Millar, Jean-François Le Maréchal and Andrée Tiberghien

Appendix 3 Practical work task: Profile Form

A Intended learning outcome (learning objective)

a	to help students identify objects and phenomena and become familiar with them	
b	to help students learn a fact (or facts)	
c	to help students learn a concept	
d	to help students learn a relationship	
e	to help students learn a theory/model	

f	to help students learn how to use a standard laboratory instrument, or to set up and use a standard piece of apparatus	
g	to help students learn how to carry out a standard procedure	
h	to help students learn how to plan an investigation to address a specific question or problem	
i	to help students learn how to process data	
j	to help students learn how to use data to support a conclusion	
k	to help students learn how to communicate the results of their work	

B1.1 What students are intended to do with objects and observables

	an observation or measuring instrument	a
use	a laboratory device or arrangement	b
	a laboratory procedure	c
present or display	an object	d
	an object	e
make	a material	f
	an event occur	g
	an object	h
observe	a material	i
	an event	j
	a quantity	k

'Mapping' the domain

B1.2 What students are intended to do with ideas

report observation(s)		a
identify a pattern		b
explore relation between	objects	c
	physical quantities (variables)	d
	objects and physical quantities	e
invent' (or 'discover') a new concept (physical quantity, or entity)		f
determine the value of a quantity which is not measured directly		g
test a prediction	from a guess	h
	from a law	i
	from a theory (or model based on a theoretical framework)	j
account for observations	in terms of a given explanation	k
	by choosing between two (or more) given explanations	l
	by proposing an explanation	m

B1.3 Objects- or Ideas-driven?

a	What the students are intended to do with ideas arises from what they are intended to do with objects	
b	What the students are intended to do with objects arises from what they are intended to do with ideas	
c	There is no clear relationship between what the students are intended to do with objects and with ideas	

B1.4 Degree of openness/closure

Aspect of practical task	Specified by teacher	Decided by teacher-student discussion	Chosen by students
Question to be addressed			
Equipment to be used			
Procedure to be followed			
Methods of handling data collected			
Interpretation of results			

B1.5 Nature of student involvement

a demonstrated by teacher; students observe	
b demonstrated by teacher; students observe and assist as directed	
c carried out by students in small groups	
d carried out by individual students	

B2.1 Duration

a very short (less than 20 minutes)	
b short (one science lesson, say, up to 80 minutes)	
c medium (2-3 science lessons)	
d long (2 weeks or more)	

B2.2 People with whom student interacts

a other students carrying out the same practical task	
b other students who have already completed the task	
c teacher	
d more advanced students (demonstrators, etc.)	
e others (technician, etc.)	

B2.3 Information given to the student on the task

a oral instructions	
b instructions on blackboard/whiteboard/OHP	
c guiding worksheet	
d textbook(s)	
e other (e.g. data book, data base, instruction manual, etc.)	

B2.4 Type of apparatus involved

a standard laboratory equipment	
b standard laboratory equipment + interface to computer	
c everyday equipment (kitchen scales, domestic materials...)	

B3.1 Nature of student's record of work on task

a	no written record	
b	notes	
c	completion of printed worksheet	
d	written account (using given structure and format)	
e	written account (free format)	

B3.2 Purpose of record

a	to assist students in learning science content or process	
b	to provide evidence that the task has been carried out	
c	as a basis for assessing the student's performance	
d	as a record which the student can use to revise for tests or examinations	
e	to help students learn how to write a scientific report	

B3.3 Audience for record

a	the student	
b	the teacher	
c	other students	
d	other	

Envisionment in practical work
Helping pupils to imagine concepts while carrying out experiments

Joan Solomon

Everyone would agree that practical work should not be the mindless following of recipes. In particular we want our pupils to see the equipment that they are using in the light of previous actions, taught scientific concepts and an active present imagination. The connection between images and actions, past or present, is intriguing and very important. It derives from the work of both Jerome Bruner and, more recently, that of Robin Hodgkin.

In the first part I will tell about some recent research in this field carried out in a small Interactive Science Centre for primary pupils, which showed the difference between simple play and intentional exploration, in several dimensions.

In the second part I will move into the domain of the school laboratory where teachers often find that everyday remembered images are too vague to incorporate potential or imaginary actions on which both explanations and predictions depend. There will be given examples from school science teaching as well as concrete 'conceptual models' which illustrate what might happen if different actions and new experiments are carried out.

Testing and explanation

Practical work in the school laboratory is an essential part of learning science. That means that it contributes to a deeper understanding of the science. The statement is so simple that seems hardly necessary to be said, and yet practical work in many British schools is in the process of losing out to a kind of testing for its own sake, and in doing so has also lost what we might call its conceptual virtue.

Far too often we find our pupils simply trying to test out which kind of

Envisionment in practical work

paper towel in the most absorbent or which piece of card is the strongest. Of course we all know that these sort of tests have to carried out commercially. The advertisements on TV every night show nappies which are being tested in just such a way. My point is that, at a basic level, this sort of activity has nothing to do with science because science is driven by the need to inquire *'Why does this happen?'* Albert Einstein once said that *'scientists were people with a passion to explain'*. Material scientists are not driven to do experiments because they want just to do some test on an unknown substance. They have an image in their heads of the structure of the substance. From this they can infer how the it might behave, and then they carry out the experiment. Such an experiment is not a quality test, and the difference is clear. It is located in the explanatory image inside the scientist's head which is driving the work. In most cases the observation and measurement carried out is coloured by some existing paradigm. This means that we need to teach our children something of this paradigm so that the activity will illustrate the model of the substance. Then what they see becomes a strengthening of some inner vision of how it 'really' is. Not only does this give the pupils' practical activity more point, it also integrates the work into the teaching and learning of science.

Let me give you two very different examples of what I mean from the history of science.

The hanging boy.

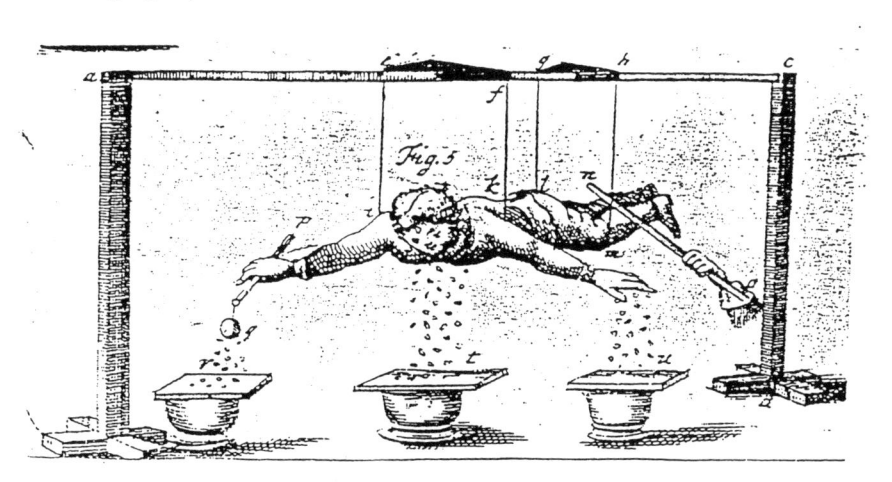

Figure 1 *From: Solomon, J. (1991) : Exploring the nature of science at Key Stage 3. Blackie and Son.Glasgow*

Joan Solomon

A demonstration rather like figure 1 was very popular in the eighteenth century. I believe that it originated in France with the Abbé Nollet, and it was taken round and demonstrated to all the 'crowned heads' in Europe. Sparks were induced to jump from the little boy's nose, and everyone was very amused! Benjamin Franklin reported that he too had seen this demonstration when he was in Europe, and no doubt he thought about it when carrying out his own experiments on electricity many years later in Pennsylvania. It was a mystery to all who saw it and, rather like the tests I spoke about before, was clearly not a practical experiment in the sense of being an exploration designed to explain a phenomenon.

A rather similar activity was being carried out here in this illustration. However this time it is an experiment. Steven Gray had noticed that sometimes the electric charge leaked away from the charged boy, and that it seemed likely to him to have something to do with the supporting ropes. He imagined the flow of electric charge up or down these ropes so he was conceptually ready to perform an experiment. He set himself a question: –

'Does the electricity really flow through the different ropes?'

He tried thick ropes and thin ropes, ropes made of hemp and ropes made of silk. (I worry about the orphan 'charity boy' suspended throughout all this experimenting!) Eventually Steven Gray wrote a long book about his results and concluded that the thickness of the ropes was not important, but the substance they were made from was important. Suspend the poor boy from silk ropes and the charges stayed in place, use hemp ropes and it did not. Reading about this early struggle to understand what was happening it is easy to see that this was the first ever experiment on 'conductors and insulators' although these words were not coined until a little later by another Frenchman, du Fay.

Ecology

This is a young science and although it is almost impossible to pinpoint where or when it all began, a good case can be for the work of Charles Elton in Wytham Wood during the 1930s.

You may see from the photograph in figure 2 that his work includes the study of birds, small rodents, butterflies and insects. And yet it was clear to Charles Elton that this was not the reality of what they were all interested in. Indeed it is said that when a young scientist applied to work with him on the animal 'populations' in the wood, they didn't get the job. For Elton the

Envisionment in practical work

Figure 2 *From: Crowcroft, P. (1991): Elton's Ecologists. University of Chicago Press. Chicago.*

wood was an ecological whole and its fluctuations of population were of huge interest just because they interconnected. In other words he had a dynamic explanatory image of the woodlands as a whole, as well as of the grasslands on its borders, which guided all the work done. It was no longer nature study with small mammal specialists, and bird specialists, working on their own. Now it was the whole habitat which was the object of thinking and experimenting.

The moral to be drawn from this for us school teachers is that nature study, or the mere counting of plants in a quadrant, is no longer appropriate practical work. Ecology demands that our students conceptualize their work in the context of the field or wood as a living interactive entity. Before they begin to explore details of the habitat the pupils need some sort of envisionment of the whole to make sense of their work.

The premise of this argument is that all good school practical work should be operating with conceptual images already existing within the pupils' mind. From this the question arises whether there is any value at all in the uncontrolled hands-on activities outside school before such concepts have been taught. Can play generate the kind of curiosity and need for explanation which could convert untaught activities into sound practical work?

Joan Solomon

Play and exploration

When practical activity is not yet guided in this way it can be *'just playing about with the apparatus'*. A visit to some of the Hands-on, interactive science centres may offer examples of this sort of activity. I want to try to convince you that an important part of this apparently mindless activity is the construction of images which are often wordless, but valuable and so closely related to action that they become in themselves also a kind of envisionment. Play can have at least two important functions to fulfil:

a) getting a feel for the equipment and the body movements associated with using it, and
b) acquiring a plan or intention for an exploration.

Ziman (1984) wrote that *'scientific experiment is never playful'*. This describes a lack of intention which distinguishes play from scientific experiment. Clearly this refers to the second function of play which is the subject of a paper, in press, written by Helen Brooke and myself which tries to map out, in detail, how play of the mindless variety, turns into purposeful exploration. The children that Helen and I observed often began rather aimlessly; they even admitted to not knowing, or even caring, about what would happen. They laughed when a car they had constructed out of cardboard flew backwards out of the wind-tunnel! Most of the adults present, their science teachers, were a little shocked by this, and clearly thought that they should have tried to make the car go down into the wind, but not so the children. The children were not concerned with the outcome. The whole activity was playful and had no purpose other than amusement. In our long term research we had to search for those magic moments when one or other of the children were overtaken by curiosity and began an investigation in which speculation about cause and effect, and the search for its corroboration took charge of their activities. Investigation is the essence of practical work at school or out of it. But it must be the kind of investigation which uses mental imaging as well as hands-on actions.

There was a classic experiment carried out many years ago by Bruner, Sylva and Genova (1974) about practice and play which deserves to be far better known by science educators. In this research there were three groups of children, two of whom were given training sessions with the sticks either to teach them how to join them together in pairs, or to demonstrate, without

Envisionment in practical work

the children being able to use the sticks how the whole extension mechanism worked, while the third group was simply allowed to play with the materials without any instruction or demonstration. Finally all the groups tried to use the sticks to reach out and get some prized object by fastening them together. We need to notice that it was the 'play' group which succeeded first. It is easy to under-rate the power of practice with a piece of equipment which gradually builds images of it in our heads. It may also build up some of these tacit images of how things work in our bodies.

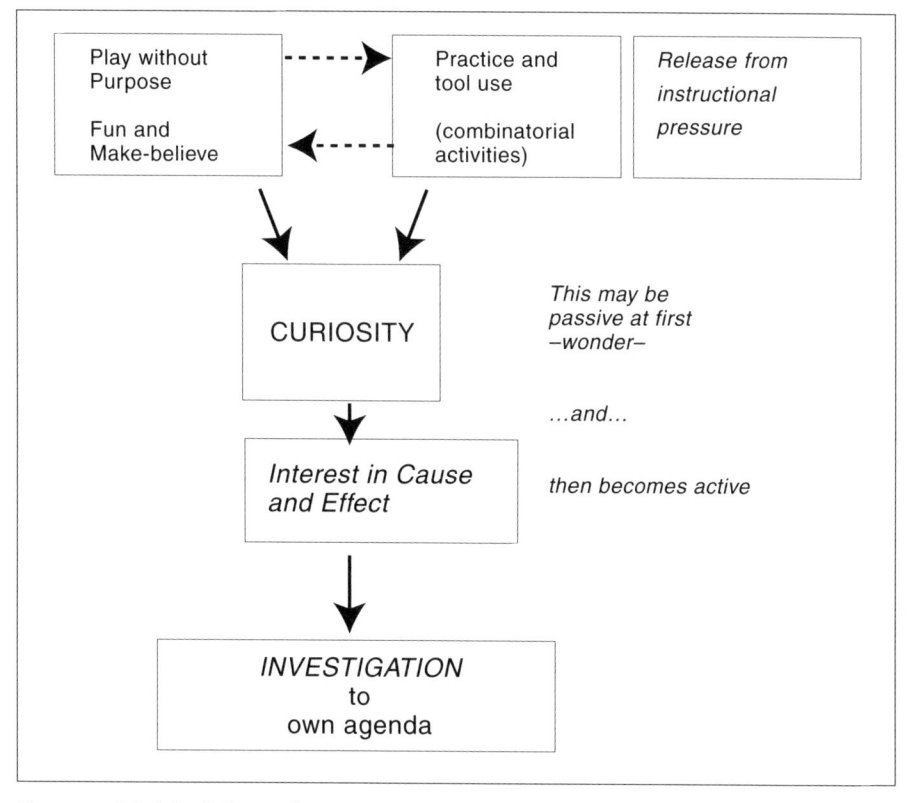

```
┌─────────────────┐       ┌─────────────────┐     ┌─────────────────┐
│ Play without    │ ----> │ Practice and    │     │ Release from    │
│ Purpose         │       │ tool use        │     │ instructional   │
│                 │       │                 │     │ pressure        │
│ Fun and         │ <---- │ (combinatorial  │     │                 │
│ Make-believe    │       │ activities)     │     │                 │
└─────────────────┘       └─────────────────┘     └─────────────────┘
           │                        │
           ▼                        ▼
        ┌───────────────┐        This may be
        │               │        passive at first
        │   CURIOSITY   │        –wonder–
        │               │
        └───────────────┘        ...and...
                │
                ▼                 then becomes active
        ┌───────────────┐
        │ Interest in Cause │
        │ and Effect    │
        └───────────────┘
                │
                ▼
        ┌───────────────┐
        │ INVESTIGATION │
        │      to       │
        │  own agenda   │
        └───────────────┘
```

Figure 3 *Model of play and investigations (Brooke and Solomon, in press)*

Joan Solomon

Making sense without words

Michael Polanyi (1958) wrote about this. For him the familiarity with tools of any kind, or new body movement, resulted in a kind of <u>body image</u> which he called *'in-dwelling tacit knowledge'*. The experienced car driver with tacit knowledge of the shape of her car, for example, hunches up her shoulders when driving between two concrete posts which might scrape the paint work, showing just how strongly the outline image of the car's chassis is indwelling in her body. Polanyi gives another example of the growth of an image which is a little closer to meaningful practical work.

When he was a medical student Polanyi was shown his first X-ray photograph during a lecture. He reports how at first neither the words of the lecturer nor the dark and light patches on the photograph made any sense to him. When sense did come it arrived as a whole. The meaning created by words interpreted the photograph, and the images retrieved from the photograph made sense of the words. The point to note here is that neither the one nor the other is primary, and that neither of them alone correspond to full internal image. This is very important in any practical work which is going to make sense to the student.

No one can write of thought and activity without paying due regard to the life work of Jean Piaget. It was one of his greatest achievements to disconnect thought from philosophical contemplation and reflection. For Piaget the function of thought was the reconstruction of action. He wrote

> The function of thought is not to contemplate but to act

Even in the famous conservation of volume experiments, in which so many jugs of orange juice have been poured out from one beaker into another, the object of the test was always the kind of thinking which involves *the imagining of action*. The child has to operate in his/her mind with reversed pouring, emptying back the juice into the first jug, or with pouring into another jug of a different shape. Only after this concrete stage can thought incorporate all these imagined actions, and supersede them. The concept of volume for the formal thinker may seem to have transcended action altogether, but it is sounder to think of it as built on the foundation of many possible actions.

Until we arrive at the far borders of modern physics almost all concepts are constructed from a whole series of possible actions. It is possible to see

Envisionment in practical work

a floating object as just that. Then it is just a simple percept. The job of the science teacher is help the student see it not just a floating in the idle way in which we all think of that word, but as in a dynamic balance between the pull of gravity and the upthrust of the water. Then a whole range of possible actions become included in what is seen, its envisionment. The act of pushing it down and the feeling of its attempt to bounce back up are incorporated into its mental image, thus making the simple percept at least half way to becoming a concept.

> (…) it is not enough to perceive the solid (object) clearly, or even to record this percept in the form of a drawing…the (symbolic) image is a pictorial anticipation of an action not yet performed…Thus what the image furnishes, to a far greater degree than perception, is a schema of action. That is why the image is more mobile and sometimes richer in content than perception. (Piaget 1956, p. 294)

Bruner appreciated this aspect of Piaget's work better than most. He wrote of children's emergence from the earlier simple world of sensori-motor action and simple percepts as

> (…) a great achievement. Images develop an autonomous status, they become great summarisers of action. (Bruner 1966 p. 13)

That is why images are so essential for doing practical work. Robin Hodgkin took these ideas forward so that the image of the object, complete with all the actions that could be performed with it, was, he claimed, somehow reproduced inside our brains.

> (…)when we visualise a pattern – a coiled rope, say – something representational of space must be happening in the deep strata of our brains, and furthermore we can <u>do</u> something with such an image if we wish. (Robin Hodgkin 1988)

Images for learning

Making sense of practical work in the school laboratory is built on these images of action. Teachers need to help students change what is seen into a vivid illustration of scientific ideas with the capacity for further action. This means that the floating object is balanced between sinking and rising, the movement of an ammeter needle indicates a flow of electric current, and the pairs of complementary muscles in the stationary model of the wing of a bird become packed with the potential movements of flight.

The teacher's role in this internal envisionment of action is rarely discussed. I still remember my own bewilderment when, as a young unreconstructed empiricist physics teacher, I was accosted by a pupil who demanded to know what her experimental result meant. *'I can see what happened'* she said, *'but you haven't explained it to us.'* Seeing, I gradually came to understand, is no more making a meaningful image to understand what is going on, than is copying down a sentence. Children need to envision heat radiation, or the flow of liquid through the turgid cells of a flower stem, while they are carrying out their science investigations. Without such images no experiments will make scientific sense of practical work. It will remain action with percepts, rather than the manipulations of action-packed concepts.

To illustrate I will take three examples.

Part I What is a shadow?

Once, long before the introduction of the British National Curriculum in primary schools, when I was beginning to teach a class of 13-14 year-colds about light, I asked them to draw the shadow of a pin-man at mid-day and in the evening. The point of this was to see if the pupils could use the newly taught idea of light travelling in straight lines related to shadows. It turned out, as we can see on figure 4 below, that some could and some could not.

Figure 4

I have to admit I was amazed! I turned to the pupil that had drawn the diagram and commented that it, *'looked more like a dead body than a shadow'*. She was not so much offended as surprised; clearly it looked quite all right to her. That was how she envisioned shadow. Then I asked all the class to write down what they thought a shadow was. Once again it was possible to divide the answers into two kinds. On one side I put all those answers where the pupils wrote about the light being *'blocked out'*, or *'not getting through'* – the ones that included action. On the other side I put those answers which just said that a shadow was a shape, an image or a 'reflection' (this is just a way of talking about a shape, and does not constitute any special alternative framework).

I found out that almost all of those who had drawn a disconnected shadow had simply written about shapes, while most of those who had drawn the shadow correctly connected to the feet of the pin-man, and also of approximately the right size compared with the orientation of the light, had given the action type of definition about light *'not getting through'*. A simple calculation showed that the writing matched well with the drawings with a coefficient of association of better than 0.76. For some of these children shadows were not just remembered (or misremembered) simple percepts. They were also complete envisionments of the action of forming a shadow.

Part II Electric current

It is often assumed that every child knows that current 'flows'. Some simple questionnaire studies carried out with colleagues (Solomon et al 1987) showed that this is one of our own misconceptions. Young pupils may only know that plugs or appliances must not be touched because they are *'electric'* or *'live'*. The action involved in this vivid and frightening envisionment is shock, pain and even death! In the pictures below the pupils were asked to put a ring round the place where there was some electricity. Because they thought of the danger of touching a plug, or any sort of 'knobby' place, we can see that quite the wrong sort of identification was produced. Electricity may have been associated with the action of 'plugging in', but that was not the sort of action which could guide the pupils towards an unfrightened use of electric current.

The teacher's role in helping pupils envision electric current in a useful way which connects with the actions they will perform in the laboratory with

Joan Solomon

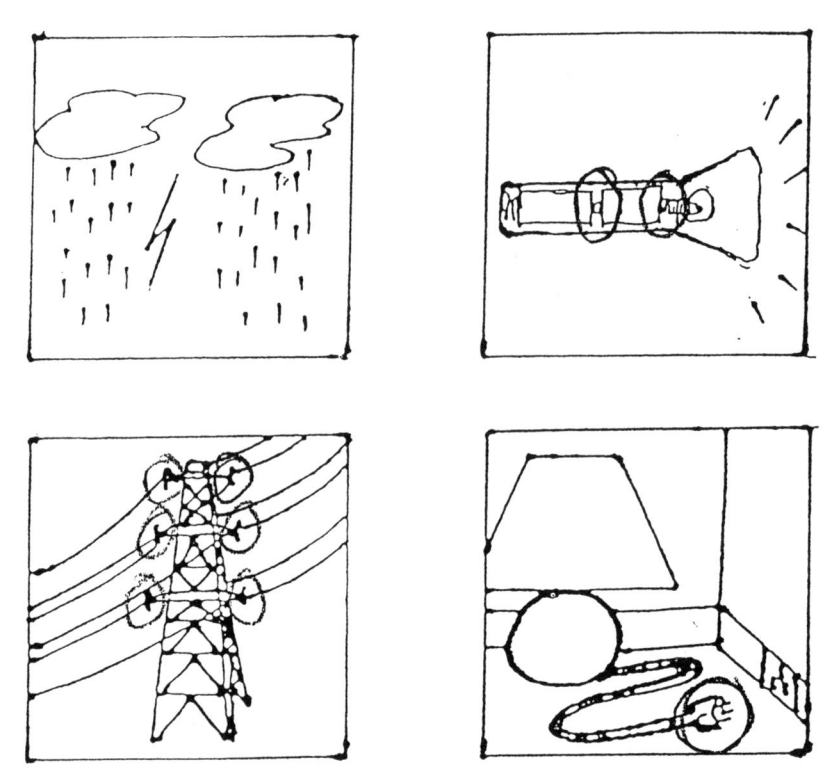

Figure 5

batteries and bulbs is quite essential. They will never actually see the current move so generations of teachers have come to the conclusion that some kind of analogy is required, and usually they tell a story of water in pipes. Several researchers have shown that this visual analogy is not always satisfactory, largely because most pupils have little understanding of fluid flow to furnish the model with the predictive power (eg. Black and Solomon 1987). So in its place we have the following concrete model, made out of a metal chain, a clear plastic tube slightly oiled on the inside nailed down on a board of wood. The children themselves put the model into action by twisting a cotton reel, to acquire a better envisionment of electric current as a stream of electrons 'chained' together so that the current flow is always the same in all parts of the circuit. It can even suggest that a thinner tube (plastic tube) will constrict and slow down the 'flow of current' (figure 6).

Questions posed to the children a week later showed that they had un-

Envisionment in practical work

ELECTRICITY   Name

This is the electric circuit   This is the model

Figure 6

derstood this model correctly and could connect the bobbles with electrons, their movement with current, and the turning of the cotton reel with the 'power' of the battery.

Part III The transpiration of plants.

This last example is included to show that students can carry out practical work which is meaningful to them, even if we might consider the hypothesis behind it as wrong or even plain silly. A class of British pupils aged about 14 were doing practical work with geranium plants based on the transpiration of water through their leaves. The teacher and her helper had spent considerable time sealing the end of the stalks to pieces of capillary tube which contained water so that it would be possible to measure the rate of transpiration. Unfortunately, as usual, it was found by the morning that only some of them were still sealed and working. That meant that many of the class had to invent their own questions about transpiration of water and to design experiments in which they measured the loss of mass of the plant+beaker system one or two days later.

Designing and carrying out their own experiments is a high risk activity, but in this case it was following a clear explanation by the teacher which

Joan Solomon

made the flow of liquid through the transport system of the plant and out through its leaves reasonably accessible to the pupils for designing their experiments.

One group of girls, of rather low achievement, decided to try out whether their plant took up more tap water or more 'natural' water. By this they meant water from the school biology pond which was green with algae but, as the girls claimed, 'natural'! This was not at all what the teacher had expected or advised the class to do, but this group of girls went ahead. As the diagram below on figure 7 shows, they had some idea of controlling variables, and of the measurements to be taken.

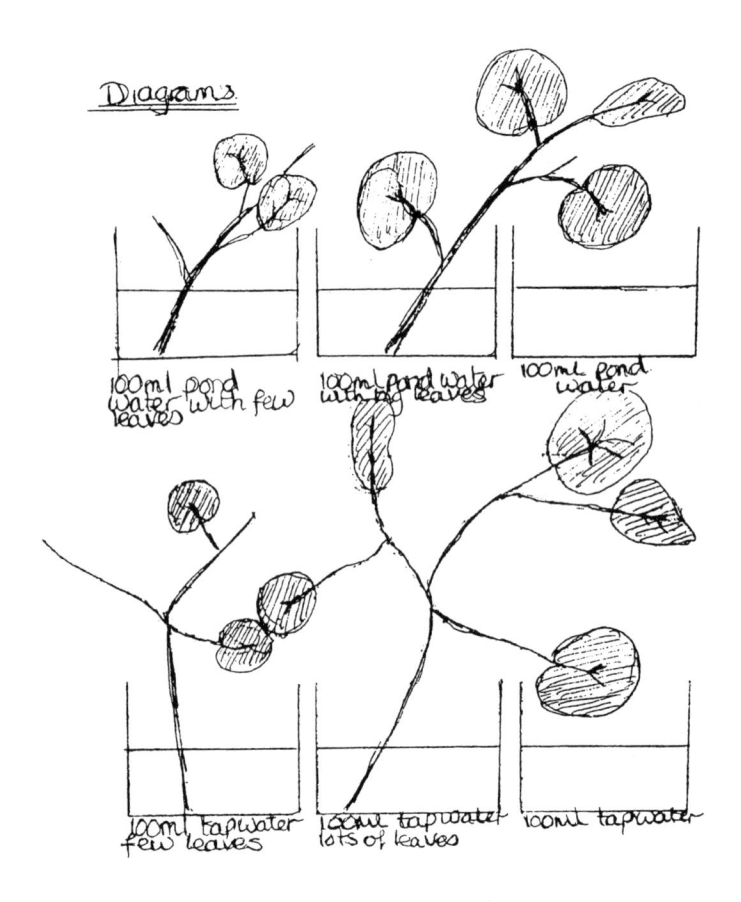

Figure 7 *Diagrams*

Envisionment in practical work

Two days later the group obtained clear results, but it was not exactly what the girls had expected. The two geranium branches placed in tap water (second row) had brought about a loss of water in the beaker. This loss was greater with the second branch with four leaves than it had been in the one with only two leaves. This pleased the girls because it conformed to their envisionment of transpiration from leaves. However the pond water had certainly not performed as expected. One of them wrote

> (…) proved my hypothesis wrong. The two (beakers of) water which had no plant in had not moved, no evaporation, also the two pond waters with plants in did not move. But the tap water did, the experiment with few leaves had gone down 3 ml, and the one with lots of leaves had gone down 6ml – doubled.

They also saw that the end of the stem which had been in pond water was soft and smelly. One of the girls wondered if there were microbes in the water that had attacked the tissue of the plant stem. These subsidiary findings helped them to see the green but 'natural' water in a new light.

It had been a valuable experiment which may well have changed the way in which these girls thought about pond water, judging by how they talked together about it afterwards. I have included this example of an experiment which refuted a naive hypothesis in this paper because it is sometimes said that to teach pupils about concepts before the experiment might bring about a dull, recipe-following activity. In this case a hypothesis about water being 'natural' was explored, and a new envisionment of the nature of pond water was produced. All of this, and the design of the experiment, was built on the secure foundation of transpiration from plant leaves.

As Piaget might have said (as much about green and rather smelly pond water, as about plant leaves)

> To know an object is to act upon it and to transform it.
>
> (Piaget 1956)

Joan Solomon

References

Black, D. and Solomon, J. (1987) The use of analogy in the teaching of electricity. *School Science Review* 249-254

Bruner, J.(1966): Toward a theory of Instruction. Cambridge, Mass. Harvard University Press.

Bruner, J. Sylva, K. and Genoa, P, (19 74) *The role of play in the problem-solving of children aged 3-5 years old*. In Bruner, J., Jolly, A and Sylva, K. (Eds.) *Play – Its role in Development and Evolution*. Penguin. Harmondsworth.

Brooke, H (1994) *Playing and learning in an Interactive Science Centre: A study of primary school Children*. Unpublished SDES Thesis. Oxford University Department of Educational Studies.

Brooke, H and Solomon, J. (in press) *From playing to investigating: research in an Interactive Science Centre for primary pupils. International Journal of Science Education*.

Gregory, R. (1986) *Hands-on Science: An introduction to the Bristol Exploratory*. Duckworth London.

Hirst. P.A. (1974) *Knowledge and the curriculum* London. Routledge and Kegan Paul.

Hodgkin, R. (1985) *Playing and Exploring*. Methuen. London.

Ogborn, J., Kress, G., Martins, I and McGillicuddy, K. (1996) *Explaining science in the classroom*. Buckingham. Open University Press.

Piaget, J.(1956 trans) *The Child's Conception of Space*. London. Routledge and Kegan Paul.

Polanyi (1958) *Personal Knowledge*. London. Routledge, Kegan Paul.

Solomon, J. (1980) *Teaching Children in the Laboratory*. Croom Helm London.

Ziman, J. ((1984) *An introduction to Science Studies*. Cambridge. Cambridge University Press.

TIMSS Performance Assessment
– a cross national comparison of practical work

Per Morten Kind

IEA's Third International Mathematics and Science Study (TIMSS) included a performance assessment component. 15 000 students from 21 countries conducted hands-on tasks in science and mathematics. This paper presents an analysis of results on the science tasks for 13-year-olds from three selected countries; England, Norway, and Portugal. The results display important differences among the countries. These differences are discussed relative to various traditions for practical work in science. It is also noticed that the results partly may be explained by the student's understanding of science investigative work, both in general and in the context of performance assessment. While students from England seem to display an awareness about what sort of responses are expected, the Norwegian and Portuguese students to a greater extend seem to have conducted the tasks as "common sense" problem solving.

Introduction

The Third International Mathematics and Science Study (TIMSS), conducted by the International Association for Evaluation of Educational Achievement (IEA), is a large scale international comparative study of student achievement. TIMSS tested students in mathematics and science at five different grades and collected contextual data from students, their teachers, and the principals of their schools. Although student achievement was measured in TIMSS primarily through written tests of mathematics and science, participating countries also had an opportunity to administer a performance assessment, which consisted of a set of practical tasks. Out of a total of 45 participating countries twenty-one carried out such a performance assessment. This paper is based on results from the performance assessment for 13-year-olds, and discusses trends for practical work in science among some selected countries. The paper also focuses on the nature of the science performance ability assessed.

By focusing on practical work under the heading of "performance assessment", TIMSS reflects a recent trend in educational assessment. This trend to a large degree is based on criticism of the extended use of multiple-choice achievement tests, that are seen to be limited in terms of capturing students' conceptual understanding and problem-solving skills and to have had a negative impact on teaching (Shepard, 1989). Multiple-choice achievement tests are also seen to assess learning outcomes in an artificial, decontextualised manner, i.e. removed from the way in which students actually learn and apply knowledge outside the classroom (Resnick and Klopfer, 1989). The new perspectives, therefore, call for more "authentic" (Wiggens, 1989) and "balanced" (Bell, Burkhardt and Swan, 1992) assessments. Students, it is argued, should be given tasks that are set in a real-world context and which require "higher level thinking or problem solving skills" (Aschbacer, 1991). Even if performance assessment appears in many school subjects, science and mathematics tend to be among the subjects where it is most frequently used. One reason for this, at least in the case of science, probably is to be found in the long tradition for emphasising practical work as a major component – a tradition that has also included practical assessment. In many ways performance assessment in science may therefore be seen as continuing an existing tradition rather than introducing something radically new.

Practical, or performance, assessment in science has very much been motivated by the aim of assessing students' ability of "doing science" as opposed to "knowing science". As displayed by the frameworks developed for such assessment this "doing" has been related to science as an inquisitive activity, with focus on investigations, experiments and problem solving. Shavleson et al. (1997) express this very clearly in stating that

> A science performance assessment comes as close as possible to putting a student in a laboratory, posing a problem, and watching as the student devises procedures for carrying out an investigation, analyzes data, draws inferences by linking data to prior knowledge, and comes to a conclusion and a problem solution. (p. 1)

Among the most influential frameworks developed to categorise the ability of doing science have been Klopfer's (1971) *"Table of specification for science education*" and "*The APU Framework for science assessment*" (APU, 1987). These frameworks typically identify certain "actions" or "behaviours" the students need to carry out in order to solve practical investigative tasks, such as "planning", "observing", "measuring", "interpreting" etc. The "scientific behaviours" have further been emphasised as skills which should be learned in science education. Performance assessment in science according to this rationale is a matter of testing if students have acquired the skills.

Critique has been raised of a too overwhelming focus on science skills and processes. Millar and Driver (1987) criticise the independent status the processes have been given and claim their rationale rests on a confusing mix of processes used by scientists, cognitive processes involved in learning science, and pedagogical processes taking place in the classroom. Woolnough (1989) criticises the tendency to teach and measure the processes independently and claim that they only give meaning in the context of the whole investigation. A reason for the contradictions may be found in a reorientation in the conception of science method since the first frameworks for practical assessment were developed in the sixties. New perspectives have influenced both the domain of philosophy of science (see Chalmers, 1982) and the domain of science education (Driver et al., 1996). Part of this reorientation arises from the criticism of a conception of science method that tends to reflect a positivist view of science (Millar and Driver, 1987, Driver and Newton, 1997). New arguments are now being brought forward in support of an emphasis on science method and science investigation in school science. A longstanding argument has been that learning science method is a way of developing the student as a problem solver. For example, reference is made in the APU Science project, to science method as "science problem solving activity" (APU, 1987). The new arguments for emphasising science methods are based on giving the student insight into the nature of science. Based on the "sociological turn" in the philosophy of science, which stresses that science knowledge is socially constructed, emphasis has been put on providing students with insight into how science knowledge becomes established (Driver et al., 1996). Science method is no longer seen basically, as set of processes, but rather as a way of arguing which leads to the common body of knowledge scientists accept. Driver and Newton (1997) claim that scientific investigative work in school science is a way of introducing students into this "scientific argumentation". Based on such perspectives learning science method is no longer a matter of learning science processes or skills, but rather of coming to understand the purpose of scientific work and learning how investigations can be carried out in a reliable and valid manner.

The discussion on the nature of science method and on rationales for including science investigations in school science is very important for performance assessment. For most countries, however, the discussion presented above is at an "advanced level". Practical work in science has been seen as a pedagogical means for teaching science content knowledge rather than a sophisticated way of "developing the mind" or learning "scientific argumentation". Accordingly, students in many countries are more commonly presented with "recipe" type of practical work rather than open-ended investigations. Performance assessment in science, including open-ended investigative tasks, is therefore a phenomenon peculiar to only a few countries. This was noticed already in the development phase of the performance assessment study in TIMSS. When the tasks were presented to the various countries taking part in the study, some of them responded critically on the "scientific" criteria implemented in the scoring guidelines; e.g. to expect measurements instead of observations, systematic control of variables etc.. It was not felt "fair" for the students to give them an open-ended task and expect this specific type of response. Also preliminary analysis of results from the pilot study indicated differences across (and within) countries in the way students approached the tasks (Kind, 1996). Sometimes students seemed to have focused mainly on giving an answer to the problem in the task, for instance "find which magnet is the stronger", and put little effort on showing that they had investigated this problem by a structured (scientific) method. Elsewhere the students had clearly put effort into presenting a large set of measurements even if this was not necessary for finding the answer to the problem. Some students, however, also presented responses that clearly indicated basic understanding of an investigation, for instance by reflecting on their choice of method and by using the gathered data when arguing for the final conclusion. Such information is important to what can be learned from the performance assessment study. It is not reasonable to interpret the results solely within a single rationale, for instance by regarding the test as measuring student's science process skills or their ability in scientific argumentation. Rather it is necessary to take into consideration how students in a particular country have interpreted the tasks assigned to them and their conception of what the task is intended to measure. Analysing the results from a such perspective gives both interesting information about various "traditions" for practical work in school science and some information useful to an understanding of science performance ability.

The method

TIMSS performance assessment included a set of 12 tasks in science and mathematics for students at age 13. The set included five science, five mathematics, and two combination tasks, integrating mathematics and science. The tasks were named according to the primary content area addressed, which for science were: Pulse, Magnets, Batteries, Rubber Band, and Solutions. The tasks were given to the students on structured response sheets. These began with a primary problem or investigation to be completed by the students, followed by a series of items (questions) worded to elicit various aspects of their investigation. The students wrote their response to each item mainly as they were completed.

The performance assessment was administered in a "circus" format in which a student completed three to five tasks by visiting three stations at which one or two of the tasks where assembled. The assignment of students to stations was determined according to a predetermined scheme. After completing a task at each station, students submitted their work booklets to the performance assessment administrator (sometimes together with products). The students were tested in groups of nine in a room with nine stations and allowed 30 minutes at each station.

A coding/scoring system was developed for the performance assessment in TIMSS, that allowed for identification of common approaches and types of error in the student response. The categories in the coding scheme, which were developed from actual student responses, were assigned a score level (0-3 points). The criteria for these scores, in the science tasks, were based on the conception that the students would use a "scientific method" (and produce a scientific report). For example: a plan would be accepted to include description of the variables it was intended to measure or observe. Data should be presented in tables (where asked for) including headings and units, measurements would score higher than just observations and many measurements would score higher than a few when identifying a trend. Coding was conducted by coders especially trained to use the TIMSS coding rubrics. Inter-coder agreement was tested and had an average of 91% across all countries (Harmon et al., 1997).

In the present analysis of the performance assessment data in TIMSS three countries, namely England, Portugal and Norway, and selected items in three tasks, Pulse, Batteries, and Solutions will be focused on. The tasks and the countries were selected in order to discuss specific aspects of practical work and practical assessment in science. The countries represent somewhat different "traditions" in practical work, a contention that is supported by their positions on the overall "scoring list" for TIMSS performance assessment (see table 1).

For each task analysed a sample of 140 -150 students from each country was taken.

Results

Country	Average percentage score on science tasks.[1]
Singapore	72 (1.8)
England*	67 (0.9)
Switzerland	65 (1.0)
Scotland	64 (1.5)
Sweden	63 (1.5)
Australia*	63 (1.1)
Czech Republic	60 (1.3)
Canada	59 (1.3)
Slovenia**	58 (1.1)
Norway	58 (0.8)
New Zealand	58 (1.5)
Netherlands*	58 (1.4)
Romania**	57 (2.0)
Spain	56 (1.5)
United States*	55 (1.4)
Iran	50 (2.8)
Cyprus	49 (1.0)
Portugal	47 (1.2)
Colombia**	42 (1.4)
International average	58 (0.3)

* Countries not fully satisfying guidelines for sampling participation rates
** Countries not meeting age/grade specifications
1 The scores are made by weighting each task equally, even though the number of items within the tasks varied.

Table 1: *Average percentage score on science tasks, TIMSS PA, 13-years old (Standard error appears in parentheses)*

Table 1 displays the overall averages in percentage score across science tasks for students aged 13 in the performance assessment. Substantial differences are seen in performance between the top and bottom, although most countries performed somewhere in the middle ranges. It can be seen that the three countries selected for a closer look, England, Norway and Portugal, are placed at very different positions. The differences in performance between one country and the next higher- and lower performing countries were rather small. The same was found for TIMSS written assessment and, with some exceptions, the relative standing of countries was somewhat similar in the two assessments (Beaton et al., 1996). The score values (percentage score) in the written and performance assessment for England, Norway and Portugal are displayed in figure 1. It can be seen that England, relative to the international average, scored much higher on performance assessment than on written assessment. A similar trend was found for Scotland and Switzerland. For Norway and Portugal the differences were very small. Both countries, however, had an opposite trend relative to the international average and scored higher on the written test. In some few countries (Czech Republic, Slovenia, and Netherlands) this trend was somewhat stronger.

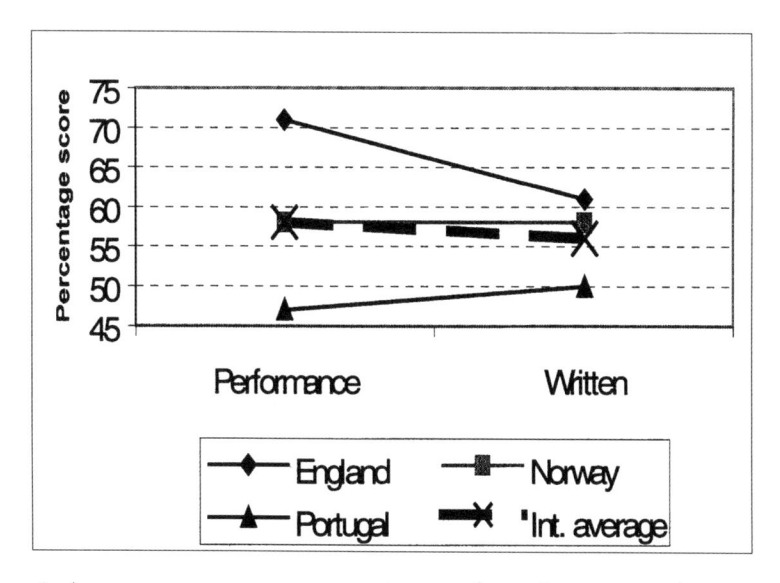

Figure 1: *Average percentages score on science tasks performance- and written assessment*

The performance assessment in TIMSS included five science tasks. The results displayed considerable variation in the difficulty of these tasks. Across countries, the Magnets task was the least difficult (international average 90%) and the Pulse task the most difficult (international average 44%). This wide range in difficulty indicates that results are best discussed according to each task rather than simply looking at the overall average. In figure 2 results for each science task for England, Norway and Portugal are presented. Because the difficulty of tasks tended to show the same pattern in all countries the results are presented as differences in percentage score relative to the international average.

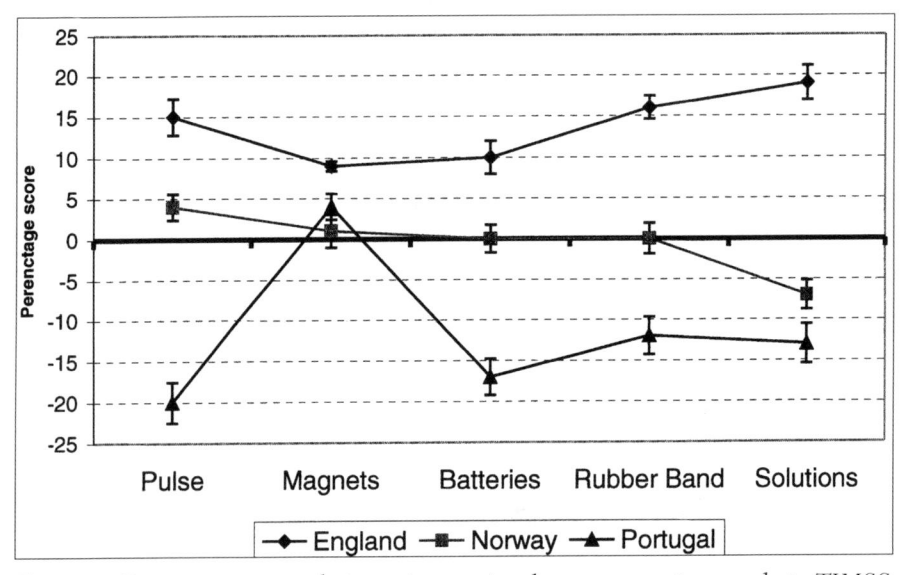

Figure 2: *Percentages score relative to international average on science tasks in TIMSS performance assessment (Standard error included for each score)*

The results clearly display the positions for the three countries. England is above the international average for every task, Norway tends to follow the average and Portugal is below. A somewhat similar pattern, however, is found for England and Norway. The difference in the pattern is most obvious for the task *Solutions*. England in this task is high above the international average while Norway is below. Portugal follows a somewhat different pattern and has a score on the Magnets tasks above the average.

The results call for a closer look at each of the tasks and some analysis

TIMSS Performance

going beyond score level. According to the perspectives presented in the introduction, it is interesting to look at students' data gathering and relate these to their conclusions. In all the science tasks the students were either asked to present data or to describe the method they used to reach the conclusion. Such data presentation normally produced some information about "the method" used by the student. Of course, this was influenced by what the student had found adequate to mention in his/her response, something which applied both to data presentation and to the student's own description of the method.

Presenting results from students' approaches to the tasks is not a simple matter. The tasks included a set of sub-items and several codes were used for categorising responses to each item. Altogether this makes a rather complicated system. In the following presentation, therefore, selected results only will be displayed. Instead of describing all variants of responses, the intention is rather to identify some main trends in each of the three countries mentioned above. Three tasks will be focused upon: Pulse, Solutions, and Batteries.

Pulse

In Pulse, the problem given to the students was *"Find out how your pulse changes when you climb up and down on a step for 5 minutes"*. The students were guided for a structured investigation through the information *"Decide how often you will take measurements starting from when you are at rest. Climb the step for about 5 minutes and measure the pulse at regular intervals"*. The first item given was *"Make a table and write down the times at which you measured your pulse and the measurements you made"*. Students' responses were coded both for the quality of the data presentation and for the quality of the data themselves. Table 2 displays, based on the latter of these, how many sets of measurements were made by the students in England, Norway and Portugal. The responses, including plausible measurements and a clear indication of their number only are included. Responses in the first column were scored 3 points, the second two points, and the third 1 point. Other responses are summed up in the two last columns.

Table 3 displays the conclusions arrived at by the students. The item given to the students was *"How did your pulse change during the exercise?"*. The table includes three types of conclusion. The first column includes responses presenting a "detailed trend", i.e. the percentage of students identi-

Country	Plausible sets of measurements			Other responses given score (1 or 2 pts)	Responses not given score (0 pt)
	5 or more (3 pts)	3 or 4 (2 pts)	2 (1 pt)		
England	34.6	21.6	15.1	14.5	14.3
Norway	21.2	28.0	11.0	12.8	26.8
Portugal	9.3	14.4	4.1	11.2	60.9

Table 2: *Percentages of students in each category for data gathering, item 1 in Pulse*

fying the number of pulse beats to increase more at first and less at the end of the exercise period. This could often be seen from students' data when they presented 4 or more measurements. The second column includes responses which just mentioned that the number of pulse beats increased as the exercise went on. The third column includes responses presenting a conclusion that obviously was not consistent with the data. The two first types of response each were given full scores (2 points), while the last was scored zero. Other types of conclusion, summed up in the fourth column, were scored 1 point.

Country	Detailed trend (2 pts)	"Pulse rate increased" (2 pts)	Not consistence (0 pt)	Other responses given score (1 pt)	Other no-score (0 pt)
England	6.3	54.3	5.7	21.6	12.2
Norway	7.3	44.5	2.2	34.1	11.8
Portugal	2.7	11.3	10.6	19.4	56.2

Table 3: *Percentages of students in each category for conclusion, item 2 in Pulse*

Taken together the results presented in the tables indicate some tendencies in the responses of students from the three countries. In both England and Norway a relatively large group of students making "many sets" of measurements (3 or more) and concluding that *"the number of pulse beats increased with the exercise"* was observed. England clearly differed from Norway with a higher number of students presenting five or more measurements. It is interesting, however, that no more students identified a "detailed trend" in England than in Norway. In a way the overall picture reveals a mismatch between the two tables. To simply identify *if* the pulse rate increased during an exercise two measurements would have been enough. It seems therefore

TIMSS Performance

as though the students have in a ritualistic way made a larger set of measurements than necessary. The reason for this, of course, could be found in the hints given in the text accompanying the task, but could also be found in the pre-learned strategy that "many measurements is important in science tasks". Some students presented data that indicated in a more obvious way that they had worked with a strategy to find *if* the pulse increases. They presented one set of measurement at the end of the exercise, or two sets, one at the beginning and one at the end. The percentage of students presenting the latter type of response is shown in the third column in table 2.

In the case of Portugal, the situation is quite different from that of the other two countries. A much lower percentage of students presenting plausible measurements and, consequently, a much higher number of students achieving "no score" (as much as 61%, see table 2) were found. The low percentage of students achieving any score at all for their conclusion suggest that the reason is not to be found merely in their ability to communicate what they really did. It simply looks as if a great number of the students had problems in handling measurements or observation of the pulse rate.

The detailed look at types of responses in the three countries gives a deeper understanding of the situation displayed for the Pulse task in figure 2. The English students achieved their high score in the task partly by following the "scientific" method of making many sets of measurements. The Norwegian students on the other hand seemed mostly to handle the technical skill of measuring pulse rate, a large group of students having presented plausible measurements in a table, but failing in the main to us the "best" method. The difference in scores between the two countries can also be attributed to other formal elements (not presented in the analysis above), such as not labelling the tables appropriately. The Portuguese students responded differently from the two other countries, and seems to have been more "confused" when presented to this task. Many of them probably have, from every day experience, the same knowledge as the students achieving full scores on the task, i.e. that the pulse rate increases when they are exercising. However, the results very clearly display that a large group of them have not been able to "investigate" this in a practical task.

Solutions

The Solutions task had some similarities with the Pulse task, but was a more open-ended investigation, with fewer hints being provided. The students were given the problem *"Investigate what effect different water temperatures have on the speed with which the tablet dissolves"*. In contrast to the Pulse task, the students were given less precise information about what method to use. The text, however, indicated that a "scientific" solution was expected by asking for a plan in the first item, and by making certain requirements for this plan. The plan should include *"what you will measure"*, *"how many measurements you will make"* and *"how you will present your measurements in a table"*. The last item asked the students to comment on their investigation: *"If you had to change your plan, describe the changes you made and why you made them, If you did not change your plan, write "no change"*. Results for the two items are displayed in tables 4 and 5 below. To achieve a full score (2 points) on the plan students had to include measurements or observations of both solution time and temperature. Responses including a "complete plan" were coded in two categories according to controlling or not controlling other variables. The percentage of students in each of these categories are shown in the two first columns. The third column includes responses not exactly specifying any plan but, more generally, describing the investigation, e.g. *"Put some cool water into a beaker. Put in a tablet. Time. Do the same with hot water"*. These responses indicate, implicitly, what variables to observe or measure. The fourth column includes responses very clearly mentioning measuring just one variable, e.g. solution time.

Country	Complete plan, other variable (2 pts)	Complete plan (2 pts)	Descriptive (1 pt)	One variable included only (0 pt)	Other score (1 or 2 pts)	Other no-score (0 pt)
England	39.8	17.8	14.6	11.7	2.8	13.7
Norway	4.2	13.9	37.1	11.6	12.7	20.5
Portugal	5.4	5.8	12.1	18.1	17.5	40.1

Table 4: *Percentages of students in each category for plan, item 1 in Solutions*

Item 5 was coded by comparing the response to item 1. If the student's plan was found complete, then a full score (2 points) was achieved by writing "No change". Otherwise the student had to suggest a change. Relevant changes (first column) were scored 2 points. Responses simply repeating the description of the method rather than evaluating the experiment were coded in a separate category (column four).

TIMSS Performance

Country	Relevant change mentioned (2 pts)	"No change", plan complete (2 pts)	"No change", plan not complete (0 pt)	Repeat information (0 pt)	Other score (1 or 2 pts)	Other no-score (0 pt)
England	18.1	37.9	24.7	2.1	4.1	13.1
Norway	4.5	9.7	57.2	0	11	17.6
Portugal	0.7	9	42.1	9.5	6.6	32.1

Table 4: *Percentages of students in each category for changing plan, item 5 in Solutions*

The results in the tables 4 and 5 show that the English students responded very differently from the Norwegian and Portuguese students. Close to 60% of the English students presented a complete plan and 39,8% included control of additional variable(s). In Norway and Portugal 18% and 11% of the students, respectively, presented a complete plan. Only a few students included control of an additional variable. As in the Pulse task, the English students could be said to follow the pattern of presenting a "scientific" response. The Norwegian students tended to (correctly) give some information about how they would solve the task, but failed to include aspects that made their plan "scientific". Thirty-seven percent of the Norwegian students presented a "descriptive" plan. The Portuguese students, as in the Pulse task, had a much higher percentage of no-score responses. Taken together the two tables tend to indicate a very different level of awareness of what is meant by a plan among students in the three countries. The English students tended to display a high awareness by presenting both good plans and by (to a larger degree than the other students) commenting upon their plan in item 5. The Norwegian and the Portuguese students tended to give an impression of not being fully aware of what sort of plan was expected from them.

Looking at the results for how the Solution task was conducted by the students reinforces the impression of different levels of awareness. As in the case of the Pulse task two tables are used to hence present "the method", based on the measurements presented by the students, and "the conclusion". The items asked were "*Carry out your test on the tablets. Make a table and record all your measurements*" (table 6) and "*According to your investigation, what effect do different water temperatures have on the speed with which a tablet dissolves?*" (table 7). In table 6 is displayed the number of measurements presented by the students. Plausible sets of measurements only are included, i.e. students including plausible measurements of both solution time and temperature Other responses are summed up in the two last columns.

Country	Plausible set(s) of measurements				Other response given score (1-3 pt)	Other response not given score (0 pt)
	>3 (3 pts)	3 (3 pts)	2 (2 pts)	1 (0 pt)		
England	31.3	9.6	46.5	0.6	8.6	3.7
Norway	16.0	15.3	19.4	3.4	38.4	7.4
Portugal	6.5	9.2	17.9	1.5	24.3	40.5

Table 6: *Percentages of students in each category for data gathering, item 2 in Solutions*

In the first column in table 7 is presented the percentage of students that mentioned a trend in the relation between temperature and solution time, e.g. *"As the water temperature went higher the tablets dissolved faster"*. The second column includes responses that focused on the effect of hot *or* cold water rather than a "trend". Other types of response are summed up in the two last columns.

Country	Trend (2 pts)	Hot or cold water (2 pts)	Other score (1-2pts)	Other no- score (0 pt)
England	83.2	3.7	3.1	10.1
Norway	22.8	34.7	10.2	31.3
Portugal	41.4	24.7	10.4	23.4

Table 7: *Percentages of students in each category for conclusion, item 3 in Solutions*

As can be seen from the tables a larger group of students in England than in the two other countries presented "scientific" responses including "many" (more than 3) sets of measurements. It is, however, interesting that a very large group of students in England (46.5%) made two sets of measurements only. This strategy tended to be linked to students' conception of temperature which elsewhere has often been described to be "dichotomous", i.e. students conceive temperature as "hot versus cold" rather than continuous (Driver et al., 1994). The students in England, therefore, seem to be divid-

TIMSS Performance

ed into two main groups according to the strategy adopted: those following their dichotomous conception of temperature and those following a strategy of making "many measurements". Table 7 indicates that students from both groups reached an appropriate conclusion. As many as 83% of the students described a "trend". In the two other countries the tendencies were quite different from those in England.

In Norway the students were more equally divided on different types of strategy, but with few students using the "best" strategy and few students using a strategy giving a zero score. Many Norwegian students seem to have used a "common sense" rather than a "scientific" method. They gathered relevant data, but did not follow the "rules" of making "many measurements" or presenting measurements for all variables. As many as 12% were found to present appropriate observations but, in a descriptive way rather than as measurements. Part of the Norwegian results, however, also include a larger percentage of students making inappropriate measurements. Seventeen percent of the sample were found to present results where solution time did not decline with higher temperature. The high percentage of students not achieving any score for their conclusion (31%) is partly due to weak measurements and a focus on irrelevant aspects.

In Portugal a similar tendency was found for the Pulse task, with low scores on data gathering. Few students presented plausible measurements including more than two measurements and 40% achieved a score of zero. However, as many as 41% correctly identified a "trend" in their conclusion and another 25% described correctly the effect of hot or cold water. This indicates that the students conducted a more meaningful investigation than indicated by their data presentation.

Batteries

In the Batteries task the students were given four batteries and a flashlight (needing two batteries) and asked to find which batteries were good and which were worn-out. They were given the hints… *"Think about how you could solve this problem" "Then work out which batteries are good and which are worn-out"*. The results are presented below for the percentage of students identifying the correct batteries (table 8, item 1) and their responses to the item 2 *"Write down how you decided which batteries were worn-out"* (table 9).

Country	All correct (2 pts)	3 correct (1 pt)	Less than 3 correct (0 pt)	Other no-score (0 pt)
England	87.6	2.6	9.7	0
Norway	91.9	0.8	2.2	5.3
Portugal	29.5	18.6	50.4	1.5

Table 8: *Percentages of students in each category for conclusion, item 1 in Batteries*

Responses for item 2 were coded according to the strategies described by the students. The first column shows the percentages of students who described a systematic strategy for testing all combinations of batteries (AB, AC, AD etc.). The second column includes responses that described other definitive strategies. The third column includes responses that described partially correct strategies, but which lack certain combinations in order to be definitive. The fourth column includes responses which described something about the task rather than the strategy used to solve it.

Country	All combinations tested (2 pts)	Not all comb., but definitive (2 pts)	Partially correct, not definitive (1 pt)	Descriptive: "I tested" (0 pt)	Other no-score (0 pt)
England	39.3	16.4	29.9	3.4	11.1
Norway	21.2	21.3	27.2	19.7	10.6
Portugal	4.8	17.1	14.6	27	36.5

Table 9: *Percentages of students in each category for describing method, item 2 in Batteries*

The results to the Batteries task supports the results obtained in the two previous tasks. Again, a higher percent of the students in England were found to present a systematic strategy than in any of the two other countries. The Norwegian students had no difficulty in identifying the batteries but, to a lesser degree expressed systematic strategies. The Portuguese students again seemed to have had both "technical" and methodical problems. Less than a third identified all the correct batteries, and very few students presented a systematic strategy.

Discussion

The results of the performance assessment in TIMSS have displayed important differences between the three countries. Some of these differences can probably be explained by general factors in the schooling system which affect students' achievement across subjects. The similar positions of the countries on the "scoring lists" for the written assessment and the performance assessment also indicate that content knowledge in science helps the students in responding to the performance assessment tasks. The results from the detailed analysis, however, clearly indicate interesting differences in practical work between the countries.

For Portugal the results tend to reflect very little experience among the students with practical work in science. The high percentages of "no score" on "data gathering" and "conclusion" indicate that many students have had "technical" problems. The extent of the problem varied from one task, and therefore one phenomenon, to another. In the Solutions task many students managed to reach the "correct" conclusion and therefore, would seem to have been able to observe or measure solution time. In the Pulse task few students at all seem to have been able to handle the basic measurement of pulse rate. For all tasks, however, the results indicate that the students have very little awareness of scientific method in that data presentation that reflecting "scientific" strategies was avoided.

For England the results contrast with those of Portugal both with respect to score level and to types of response. In all tasks a high percentage of students presented plausible measurements and tended to display a high awareness about using a *scientific method*. The students also scored relatively higher on the most open-ended task (the Solutions task), as compared to the international average. The results are in accordance with the role of practical work in science education in England. The National Curriculum for science in England and Wales defines *"Experimental and Investigative Science"* as one of four "attainment targets", and this is assigned a 50% weighting the first years in primary school and 25% later in lower secondary (Duggan and Gott, 1995). Other studies have also shown that students spend between 40 to 80% of their time in science classrooms on practical work (Beaty and Woolnough, 1982). The English results, however, may be related as much to the type of practical work as to the quantity. England has been a guiding nation in implementing "processes of science" as presented in the introduction to this article. Studies like the APU Science, and of course the National Curriculum and the National Assessment, have placed a strong focus on science method and investigative work. The types of tasks involved in the performance assessment in TIMSS fit into the paradigm for this implementation. It may therefore be that students in England have met similar tasks and have achieved an understanding of how they should be handled.

The Norwegian results lie somewhere in between Portugal and England. Norwegian students seem to handle the technical skills very well, but use less structured and "scientific" methods than English students. The responses indicated that the students sometimes focused on "finding the answer" rather than putting effort into presenting a "good method". In contrast with England, the students seemed to handle the cued tasks, with many hints given, relatively better than the most open-ended investigative tasks. The Norwegian results also make sense according to the place and role of practical work in school science. Practical work is highly favoured among teachers, but without the same attention being given towards learning science investigation as is found in England. Among the goals given most priority in practical work are "motivation for learning science" and "learning science content knowledge"(Lie, Kjærnsli og Brekke, 1997). It is also found that teachers mostly use "recipe" types of practical work guiding students towards certain observations and experiences (ibid.).

The results within Portugal, Norway, and England probably have parallels in other countries. The English version, with emphasis on science method, is a rationale to be found also in other Anglo-American countries. In the case of Scotland this seems to have favoured the students in the same way as for England. The US version is different. They too have a tradition both for emphasising science method and using performance assessment, but their students still score low on the performance assessment test in TIMSS. Other countries follow more closely the example of Norway, with focus on practical work mainly as a pedagogical means. Many similarities are found between Norway and Sweden, but still the Swedish students are closer to Scotland in their total score. All in all the results clearly indicate the need for going beyond a one-dimensional scale when analysing practical work from an international perspective.

The different traditions for practical work in Norway and England are also of interest with regard to the development of the rationales presented for science investigative work in the introduction. Two somewhat different perspectives were presented according to "skills and processes" and "scientific argumentation". The former focused on developing certain scientific problem solving skills, the latter was more concerned with understanding the nature and purpose of science method. From the perspective of "skills and processes", each of the items in the performance assessment tasks may be seen to measure specific skills such as, "planning", "data gathering" and "interpreting". From the perspective of "scientific argumentation" the items together may be seen to measure student's ability to present a conclusion based on reliable data. This could further be seen to reflect students' understanding of the purpose of science investigation. It is interesting then to reflect on what perspective has guided *students* when carrying out the performance assessment tasks. An aspect to notice, in this perspective, is that all tasks in the performance assessment included rather simple "problems" which, ultimately do not demand structured methods in order to arrive at an answer. For instance, it is possible to find which magnet is stronger just by "playing" with the magnets. It is also possible to find that pulse rate increases without making any measurements. The problems, therefore, could be solved by more "common sense" methods than the "scientific" methods.

The Norwegian students, as seen by the results, indicate that they have sometimes focused on this *common sense* type of problem solving. They seem not to be fully aware of the scoring criteria or the "rules" for doing performance assessment. For instance by writing "I tested" instead of presenting the exact method they used for testing in the Batteries task. From a such perspective it would seem that measurements are included partly because they are needed for solving the problem, but partly also "just because they are asked for". The lack of relationship between data presentation and conclusion, supports the latter conclusion. It is, however, also likely that this reflects a lack of understanding of the purpose of science investigations.

In contrast, the English students seem strongly to have learned "the rules". They seem very aware of what sort of responses are expected. Questions may still, however, be asked about their rationale for conducting the tasks. Some details in their results also indicate "ritualistic" data gathering and the lack of a relation between data and conclusion. It may therefore be that the English students also are not fully aware of the purpose of doing scientific investigations. It may be that many of them have just learned certain strategies to involve in scientific investigative tasks, without really understanding why these elements are involved.

If science investigative work is to be included as a major topic in science education, it must be handled meaningfully. A rationale that presents science method simply as a "useful tool" for problem solving should not be supported. Both teaching sequences and assessment seem to have developed a common understanding that following certain ritualistic strategies *is* problem solving. The assessment, such as found in TIMSS, tends to reveal a "play" where students ritualistically present these strategies when solving the tasks and achieve full score for their method. A way forward could be to put more focus on the purpose of scientific investigations both in teaching and assessment. Teaching "standard rituals" must be avoided and greater attention paid to discussing questions like: What would be an appropriate method in order to collect reliable data?, and, How may the data be interpreted?. Students should be trained to argue for an interpretation in the light of evidence that is available. Assessment tasks according to this should probably not include as many hints as are to be found in TIMSS. Further development, therefore, is needed to find an assessment format that demands that students more uniquely display their ability in "scientific argumentation".

References

APU (1987). *Science report for teachers: 9. Assessing Investigations at Ages 13 and 15.* Department of Education and Science. Welsh Office and Department of Education for Northern Ireland.

Aschbacer, P.R. (1991). Performance Assessment: State Activity, Interest, and Concerns. *Applied Measurement in Education*, 4 (4), pp. 275-288.

Beaton, A.E. et al. (1996). *Science Achievement in the Middle School Years.* IEA's Third International Mathematics and Science Study (TIMSS). TIMSS International Study Center, Boston College

Beatty, T.W. and Woolnough, B.E. (1982). Why do practical work in 11-13 science? *School Science Review*, 63, 225, pp. 768-70.

Bell, A., Burkhardt, H. and Swan, M. (1992). Balanced Assessment of Mathematical Performance. In Lesh, R and Lamon, S.O. (eds)., *Assessment of Authentic Performance in School Mathematics.* Washington D.C., AAAS Press.

Chalmers A.F. (1982). *What is this thing called science?* Milton Keynes, The Open University Press.

Driver, R. and Newton, P. (1997). *Establishing the norms of scientific argumentation in classrooms.* Paper prepared for the ESERA conference 2.-6. Sept. 1997. Rome

Driver, R., Leach, J., Millar, R., and Scott, P. (1996). *Young People's Images of science.* Buckingham, Open University Press.

Driver, R., Squires, A. Rushworth, P. And Wood-Robinson (1994). *Making sense of secondary science, research into children's ideas*. London, Routledge.

Duggan, S. And Gott, R. (1995), The place of investigations in practical work in the UK National Curriculum for science. *International Journal for Science Education*, vol 17, no 2, pp. 137-147.

Harmon, M. et al. (1997). *Performance Assessment in IEA's Third International Mathematics and Science Study*. TIMSS International Study Center, Boston

Kind, P.M., (1996). *Exploring Performance Assessment in Science*. Doctorate dissertation submitted to the University of Oslo, March, 1996.

Klopfer, L.E. (1971). Evaluation of learning in science. In B.S. Bloom, J.T. Hastings, and G.F. Madaus (eds.), *Handbook on formative and summative evaluation of student learning*, pp. 559-642. New York, McGraw Hill.

Lie, S., Kjærnsli, M. and Brekke, G. (1997). *Hva i all verden skjer i realfagene? Internasjonalt lys på trettenåringers kunnskaper, holdninger og undervisning i norsk skole*. ILS, University of Oslo.

Millar, R. and Driver, R. (1987). Beyond Processes. In *Studies in Science Education*, 14, pp. 33-60.

Resnick, L.B and Klopfer, L.E. (1989) .*Toward the thinking curriculum: current cognitive research*, Yearbook for Association for Supervision and Curriculum Development

Shavelson, R. J., Solano-Flores, G. And Ruiz-Primo, M.A. (1997). *Towards A Science Performance Assessment Technology*. Paper prepared for the 7 th EARLI Conference, Athens.

Shepard, L.A. (1989). Why we need better assessment. *Educational Leadership*, 46 (7), pp. 4-9.

Wiggens, G. (1989). A true test: Toward more authentic and equitable assessment. *Phi Delta Kappa 70 (9)*. pp. 703-713.

Woolnough, B.E. (1989). Towards a more holistic view of processes in science education. In Wellington, J. (ed). *Skills and Processes in Science Education*. London, Routledge.

Touched by a disgusting fish

Dissecting squid in biology lessons in a comprehensive school

Piotr Szybek

The objective of this article is to interpret an event: the dissection of squid in biology lessons in a comprehensive school. The interpretation was based on an analysis of interview transcripts which highlighted potentials emerging from action. The event was enacted on a stage, constructed by the dissection. The meaning of dissecting was a "nice arrangement" of the squid (the term is derived from an expression used by one of the interviewed pupils), thus not being seen as "disgusting fish" but becoming something that could be touched. Personages appearing on the stage could assume two different roles, according to gender: "tough boys" and "sissy girls".

Introduction

This article describes the dissection of squid by pupils in a Swedish comprehensive school (about 14 years old). The background is a study sponsored by the Swedish National School Agency in 1994/95 and performed by the author, then a teacher of biology. The initial interest was the author being puzzled by vociferous demands, voiced exclusively by boys, worded e.g. "when are we going to cut up squid" (not: "are we etc."). At the same time it did not appear that dissecting squid really contributed to mastering some relevant subject matter. The objective was thus to find out "what are they really up to when they say they want to dissect?".

Rather than aiming at proposing new ways of dissecting, or of using dissection to promote inculcation of scientific concepts, the paper gives a description from the point of view of curriculum theory, looking for potentials implied by the activity. These potentials can concern equity and the way scientific knowledge is used in social and political contexts. Following Meyer-

Drawe (1984), bodily aspects of human existence are stressed, which grounds the discussion in concrete experiences, and provides links between the "cognitive" and the "affective".

Touch as event

In this article an attitude is maintained which phenomenologists call "rejecting the natural attitude". Schütz and Luckmann characterise the *natural attitude* as assuming that the world can only be understood in one way, the one currently adopted (Schutz & Luckmann 1977). An established world-view can be challenged, refuted, and exchanged for another, which in its turn becomes *the* established world-view, assumed to be the only possible. In day-to-day practices this must be so: Actions must be based on a confidence in the order of things. In some practices, however, abandoning the natural attitude can be fruitful. I wish to propose that evaluation and educational research are among these.

The way to do this is to disinterest oneself in the question "how things are". Instead, the question "as what do things appear" comes to the fore. This is the attitude of the phenomenological epoché. (Husserl 1976, 1977, 1980, 1989, Giorgi 1975, Karlsson 1993, Moustakas 1994, Sages et al 1999, Sages & Szybek in press).

"A is touching/touched by B" means that objects have certain ways of appearing (modalities of appearing) which permit us to see these objects as touched. This is what Ricoeur (1992) calls interpretation. To interpret something is to make explicit what objects appear, how they are connected with one another, and in which modalities they appear. Interpretation is thus making explicit meaning within a range of possibilities.

This can be expressed less technically, by using the metaphor of the stage.

In a South-Indian Kathakali theatre, staging a play based on the Ramayana, the actor playing the demon Rawana makes a large step. The recitator chanting the text informs us that Rawana has just left India and is now arriving in Sri Lanka. The words of the recitator, and the movements of the (miming) actor are *shaping the stage*, making it:

- contain India and Sri Lanka within a space which we perceive as being at the most 20 square metres

Piotr Szybek

- be measured by distances which are not uniform – on one occasion a step covers the distance between India and Sri Lanka, on another it is a normal 0.75 m step.
- feature beings like demons and gods. An important trait of a stage is *what kind of entities* are expected, as opposed to making us start. The appearance of a man with six-shooters, boots with spurs and a Stetson hat on a *Kathakali* stage would not be expected.

The example shows that a stage is constructed by words and actions.

Two levels of events

On one level the event is that a girl is sitting and touching a squid's tentacles for twenty minutes with various tools. Expectations, characteristic for a given stage, determine relevance: we may decide that the fact that the pupil is, during those twenty minutes, changing her/his way of sitting on her chair is not relevant, and can be dismissed as event, but that the expression on her face while she is touching the squid is relevant and must be carefully reported. We may also decide that all this is irrelevant, and that the event is "the pupil has not been dissecting in a proper way, but only played with the tentacles". The two decisions construct different stages, where different events are enacted. One of those who enact events, and who can be expected to act in different ways depending on which stage s/he decides for, is the teacher.

On another level the touching of tentacles for twenty minutes becomes an event of another kind. Hannah Arendt (1969) is speaking for the whole phenomenological tradition when she defines "events" as "occurrences that interrupt routine processes and routine procedures". She thus characterises the stage of human existence, a stage of unique and unpredictable events. According to Arendt (ibid.), predictions tell us about what can happen if we do not act. Authentic action, on the other hand, is not possible to predict. (A recent, and spectacular, example is the fall of communism in Central Europe, which took everybody by surprise – except the immediately involved, who were not making predictions, but simply acted towards a goal which was in accord with their values.)

Predicting what might happen during a lesson, to "interrupt routine processes and routine procedures", is not possible; that is, if there is room for the stage of authentic human existence to be introduced. Instead, a range of possibilities can be described, circumscribing the ways in which things can be expected to appear. Naturally, this gives humans the possibility to change the state of affairs, and to introduce new possibilities, and with them, new ways of appearing to be expected. Humans thus construct a stage of events.

It is claimed in this paper that "touching/being touched by", which occurred between the pupils and the squid was an event of this kind. To see it as such, it is necessary to adopt the attitude of the phenomenological epoché.

Methods of gathering and processing data

The material was videotapes and interview recordings, made by myself. It was obtained in a comprehensive school in Sweden, in the 1990s. The interviews were conducted by myself, at the school, in free periods during school-days. Interviews started with a general reference to the dissection. I then followed the interviewee along the path s/he indicated with her/his answer: the question about how it was to dissect the squid gave the pupils the opportunity to talk about things being disgusting or not, and being able to stand it or not. The pupil here made a connection to having eaten squid before. This lead me to ask about eating other animals. The methodological underpinning of this is the phenomenological idea of the *horizon of expectations* (range of possibilities, Husserl 1976; 1977). I was trying to follow the interviewee into this range, and then explore it by providing questions which were variations of the theme I could see in her/his answers (Husserl 1977; Giorgi 1975).

Interview transcripts were analysed using the phenomenological method described in Sages et al (1999) and Sages & Szybek (in press). In this method, one proceeds by dividing the text into short units, and then examining every unit to identify (1) objects, (2) their predicates, and (3) the modalities in which objects appear.

The modalities of interest here were:

1. function: perceptive, imaginary, signitive. The interviewee referred correspondingly to objects being there before one's eyes ("the squid we dissected"), imagined objects ("the cat we would be dissecting"), or "generic" objects ("a squid or a cat in general").
2. affect and volition.

Piotr Szybek

At least three more modalities can be discerned, but they are omitted here, as having no bearing on the discussion. To give an example, an excerpt of the interview with Pauline is analysed:

> Pauline: Yes we started looking inside it so then I didn't think it was disgusting but it was that you were like plucking with the pincer then you noticed it was no good trying that you should just go at it or you would never get anything done and you had to touch and feel it or you never would get anything done, just sit and smell it after it I thought they were very funny

The result is shown in figure 1. The modality of affect and volition is shown, and of objects, the acting persons. This becomes a foundation for interpreting the material, in this case as follows:

> Dissection is for Pauline a quite unpleasant task which has to be got over with, the alternative being still worse. She is not engaging wholeheartedly in it. The way of speaking of herself (as "you" or as "I") indicates the distance she puts between herself and the dissection – it is "one" who is "going at it" and "getting it done", and not "I", and it is "you" who is forced ("had to") "touch and feel it" and who would "sit and smell". Once the task is "got done", she is free to engage more wholly in the situation – she speaks about "I" thinking of it as "very funny.

This interpretation is made possible by unveiling the way in which the objects "you" and "I" are connected to other objects and how they are modalised.

Figure 2 shows what the unveiling of other objects and relations can imply. The interpretation made possible here can be:

> The description is moving from the stage where the dissected squid is held at some distance from the dissecting person (by not naming specific objects, but using pronouns only, and not naming the activity) to a stage where the person is in a closer contact with the squid (activities are named, which describe the person as tangibly acting). The person present on the "close contact" stage is, however, an impersonal "you", the specific "I" appears on the "distance" stage. "I" and "you" are thus involved in different kinds of event. This event is touch, and it is a different touch on the two different stages.

Touched by a disgusting fish

objects	modalities of affect and volition (modality indicators in italics)				
acting persons	negative	faintly negative	void	neutral (possibility of negative exists)	positive
somebody, I			(somebody) looks inside		
you				I didnt think it was *disgusting*	
you			you were plucking with a pincer, like		
you		then you noticed it was *no good trying*			
you					you *should* just *go at it* or you would never get anything done
you		you *had* to touch and feel it			
you		*or* you never would get anything done, *just sit and smell it*			
I					after it I thought they were *very funny*

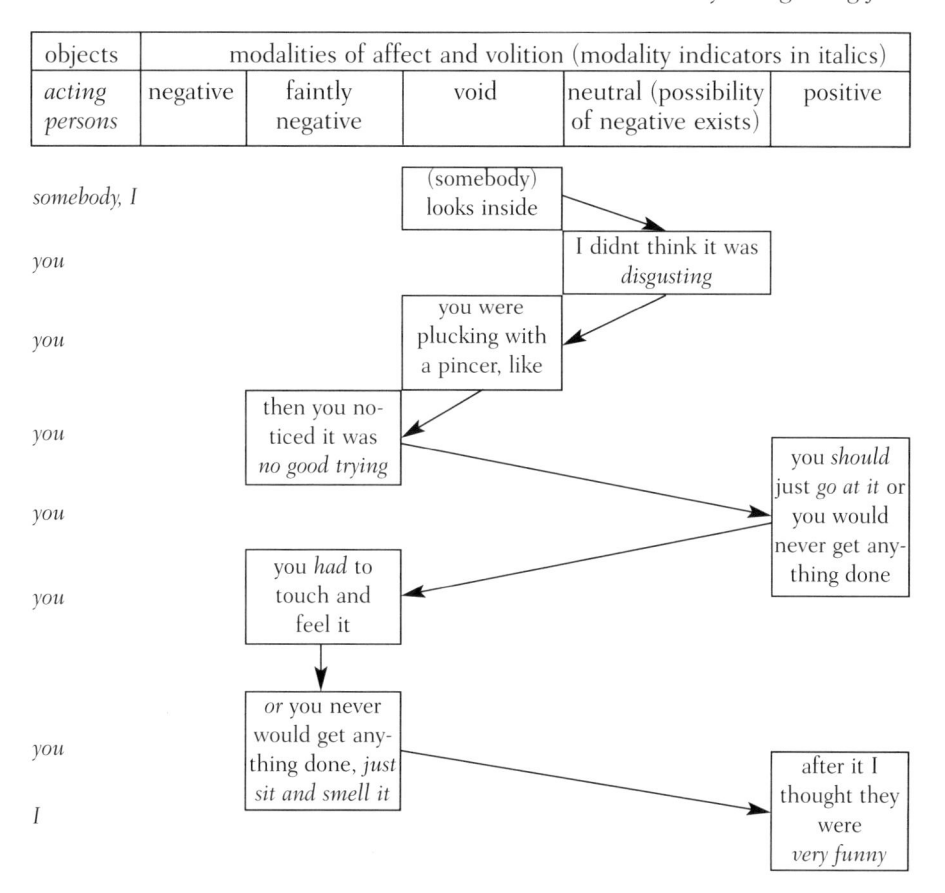

Figure 1. *Pauline talks about "going at it": Objects and modalities in an interview excerpt*

Objectivity

In the epochéattitude things appear differently than in the natural attitude. An example is "fright", connected to killing the fish Veronique caught. In the natural attitude the question asked would be "Was Veronique really frightened by the prospect of killing the fish?". In the epochéattitude the relevant question is "In which way does 'fright' appear, of which she is speaking". An important consequence of it is that "fright" becomes linked to what Veronique is doing and experiencing in a given situation (here: the prospect of killing the fish). This enables a precise way of speaking about "feelings" and linking them reliably to observable occurrences, i.e. to what given peo-

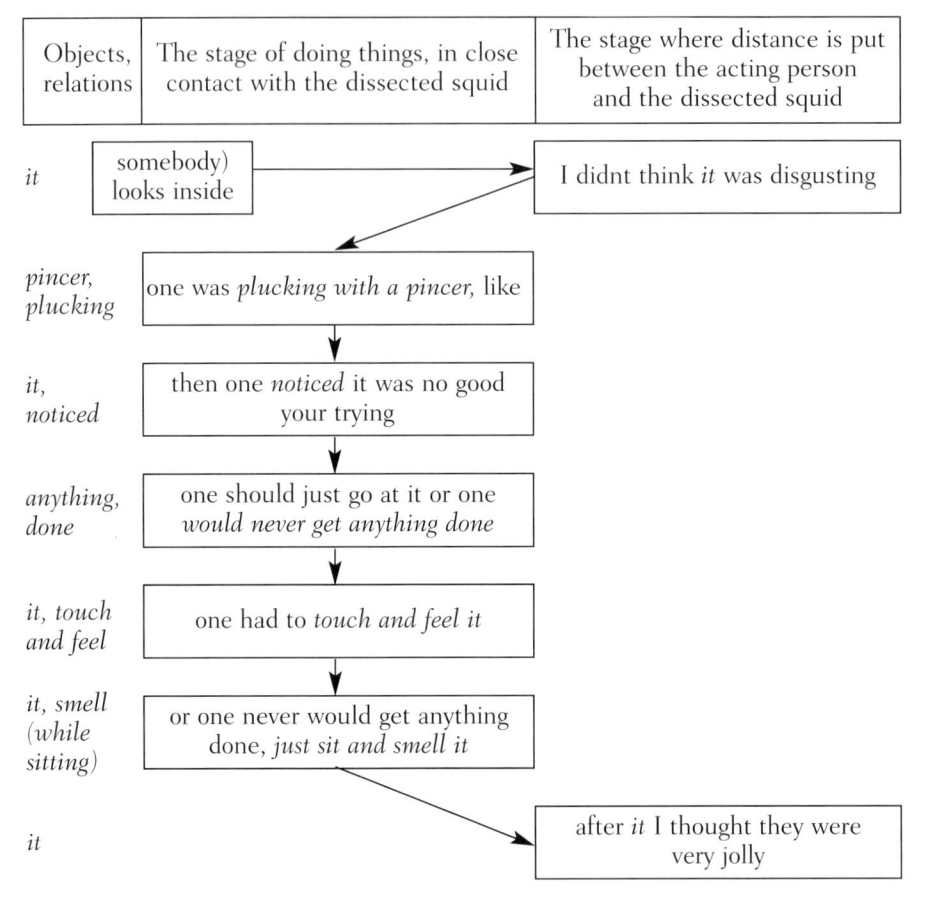

Objects, relations	The stage of doing things, in close contact with the dissected squid	The stage where distance is put between the acting person and the dissected squid
it	somebody) looks inside	I didnt think *it* was disgusting
pincer, plucking	one was *plucking with a pincer*, like	
it, noticed	then one *noticed* it was no good your trying	
anything, done	one should just go at it or one *would never get anything done*	
it, touch and feel	one had to *touch and feel it*	
it, smell (while sitting)	or one never would get anything done, *just sit and smell it*	
it		after *it* I thought they were very jolly

Figure 2. *Pauline talks about "going at it": Choice of pronoun constructs different stages*

ple are doing and experiencing in given situations. Thus, when we speak about "feelings", "affects", "emotions" we mean affective modalities of experiencing objects. This refers to a given situation ("when Veronique hooked a big fish").

Objectivity is, in phenomenology, connected to experiencing by subjects. Thus I asked Veronique and Bert if they would dissect a cat. Their answers constituted different objects. One could say that the cat discussed with Veronique was another cat than the cat discussed with Bert. When Veronique declares she would not dissect a cat, she makes a cat appear, which has the quality of "non-dissectable", whereas Bert, by saying he would dissect a cat if that was done "as a lesson like" makes it appear as "dissectable" (see figure 3).

Validity

The question "is this what the pupils *really* thought?" is not legitimate in the phenomenological framework. I would like to submit that this is in accord with the educational researcher's task: to explore conditions of co-operation for which a knowledge of the range of possibilities is most expedient.

Construct validity has been treated by Lee Cronbach (1971; 1989) who devised the concept of *nomological networks* into which the construct should fit. Knowledge about "touch" and "nice arrangement" would thus be valid, if we could show patterns of causality in which these constructs function.

This idea resembles the treatment constructs receive there in natural science. Acceleration is considered a valid construct there because coherent relationships can be formulated between it and other entities. Most important, however, is the possibility to demonstrate and corroborate these relationships by experiments resulting from the construct being based on corporeal experiences. In the case of constructs based on what people tell the researcher no observations comparable to physical experiments are possible. The phenomenological method grounds constructs in the *network of corporeally based experiences* described by the objects experienced by persons and the modalities of experiencing.

There is a limitation to this, which can be seen as paralleling the limitation imposed by apparatus in natural science research. A description of apparatus enables the readers to detect flaws in it which lead to flawed results; consequently, a better apparatus is designed, giving improved results. Here, interviews were analysed by an "apparatus" described exhaustively in Sages et al (1999) and Sages & Szybek (in press).

A teacher or educator may lack the time and opportunity to undertake such tests. There is, however, the test of practice: the use of a construct in teaching and teacher training can change the practice. The practitioners themselves will be able to assess the scope of this change and its direction, toward desirable or undesirable results. Here, the limitation is the nature of the practice, which may not permit a change of outlook necessary to put a new construct to use.

Piotr Szybek

Results

Results are presented on two levels: (1) of the straightforward narrative, and (2) of a phenomenological analysis of texts (interviews).

In the natural attitude:

I had quite recently transferred to the school in question. In the school where I worked prior to this, squid were not dissected. I used to dissect shrimps and herrings and the rationale of doing this was a possibility to compare invertebrates and vertebrates, and pointing out adaptations to life in marine environments. Dissecting squid, in my eyes, lacked this, and I could not see what pupils could learn from it. However, pupils urged me vociferously to dissect squid. That was something which had been done in that school. Peculiarly enough only boys urged, girls being silent on the subject. I gave in and performed the dissection. The impression I got was that the pupils were not doing biology. I decided to study what it was they were doing. I videotaped some lessons. A résumé of one of them follows:

The lesson begins by the teacher (a colleague of mine) giving instructions: to fetch squid and dissecting tools from the teacher's desk, and to detach parts, which are to be displayed according to a pattern. The pupils proceed according to the instructions. I stand first between the window and the teacher's desk and film Veronique and Beth. Beth is working according to the instructions, whereas Veronique keeps picking at the tentacles with a pincer and a dissection-needle, lifting them up etc.

Bert and Jake are carrying on a little noisily. I move to a position about two meters from them. This attracts Jake's attention, and he shows me the piece of squid he is working on and says "fillet". Bert then holds up another piece of squid and shouts "Beth!". Beth looks round and shrugs her shoulders. I ask: "What are you showing to Beth?" and Bert answers "Guts... er... testicles". Then he takes the piece of squid and goes with it to the other end of the classroom, where the sink is, then out in the corridor, and comes back after a few minutes. The pupils at the desk in front of Bert's and Jake's are showing me their arrangement. The teacher now gives the instruction to finish the dissection, to tidy up, and to return the tools.

Veronique and Bert were placed, partly by myself, in the focus. I wanted to know more about what they were actually doing. I interviewed both these pupils, and three pupils from another class (in the same form), who also have been dissecting squid. The interviews were subsequently processed as described above. Looking back at the events filmed during the lessons results in the story re-told in the phenomenological attitude.

Pupils were given a questionnaire (n = 47), where they were asked (among other things) if they
 1. liked dissecting the squid
 2. would dissect a cat

Boys tended to answer affirmatively to both questions, girls negatively. A χ^2 test gave a significant difference ($p < 0.001$).

In the attitude of the phenomenological epoché.

The analysis of the interview material showed a number of objects. The way these objects appear is shown in figure 3.

* *The squid*
appeared as

1. disgusting, slimy (feeling softer and slimier than a human body)
2. equipped with tentacles; for Veronique these were "legs" and when talking of them she performed a sort of a *dance,* with her arms.
3. a fish
4. an object of dissection. For a modalisation of this, see figures 1 and 2.
5. potential food

* Various animals, *a rabbit, a cat, a dog, a cow, a bear, a cod, a chicken* and *a half pig.*
Some of them were linked to actual occurrences in the past. Others were linked to imaginary, "virtual", occurrences directly connected with dissecting, and directly connected to eating, thus having the indirect connection to dissections, established during the interviews. This made all these animals *dissectable or not.*

Bert was asked if he would dissect various animals, among them a rabbit, a cat and a dog. Pauline named the "half pig" herself, I asked her about the chicken. Veronique and Pauline were also asked to imagine a situation where they are married and their husbands hunt (a traditional occupation in Swedish farmers' lives), and bring home a rabbit or a part of a bear.

object, predicate	related to	function modality	affective/ volitional modality	pupil
squid, undulating with tentacles	1. the squid that swims in the sea 2. the dissected squid	imaginary/ perceptual	enjoyment: expressed by Veronique moving her arms in a sort of undulating dance, when refererring to tentacles (as "legs")	Veronique
dog, dissected	dog, at home squid (by dissection in classroom)	imaginary	"difficult" to dissect, but possible ("as a lesson, like")	Bert
cat, dissected	squid (by dissection in classroom)	imaginary	dissectable	Bert
cat, dissected	squid (by dissection in classroom)	imaginary	non-dissectable	Veronique, Priscilla
dog, dissected	squid (by dissection in classroom)	imaginary	non-dissectable	Veronique, Priscilla
pig, half	squid (by being potential food) pigs (whole) P. "knows as filthy", and pigs she thinks are "cute".	imaginary/ perceptual	non-dissectable (dissection delegated), possible to transform into meat	Priscilla
meat, eaten	half pig (and squid, see half pig)	imaginary/ perceptual	possible to eat, if she does not think of it having been pig	Priscilla
pig, half	squid (by being potential food)	imaginary	non-dissectable and impossible to handle, unless "nicely arranged"	Veronique
meat, "nicely arranged"	half pig (and squid, see half pig)	imaginary/ perceptual	possible to handle and eat, because of being "nicely arranged"	Veronique
bear, dead	squid (by being potential food) teddybear P. had as younger	imaginary	non-dissectable (dissection delegated), possible to transform into meat	Priscilla
chicken, dead	squid (by being potential food)	imaginary	impossible to handle ("If you said: pluck it, I would refuse")	Priscilla
codfish, caught by Veronique	squid (by being potential food and by "cutting up")	imaginary/ perceptual	enjoyment: "oh, I have a bite" fear connected to "breaking off its head"),	Veronique

Figure 3. *Objects appearing in the interviews*

Touched by a disgusting fish

```
Veronique: I can't kill the fish, like, cause you have to sort of break its head off.

         I can't kill the fish like         cause you have to sort of
                                             break its head off.

specified                                                              unspecified
                    ( I ) ═══════════════ ( you )

                          "cannot":
                   can/cannot ─────── have to...
                          because...

unspecified        kill.. how?   condition    break off...              specified
                   ... like      of killing   what?...head

                   whom?                      whose?
                  (the fish) ─── identity ─── ( its )
```

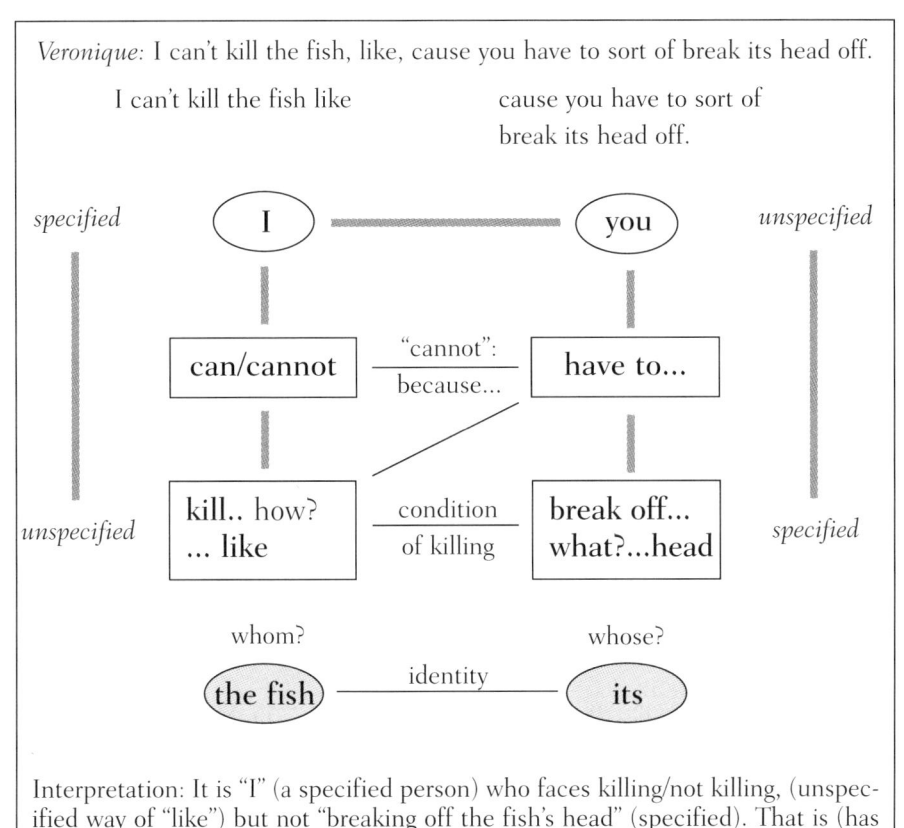

Interpretation: It is "I" (a specified person) who faces killing/not killing, (unspecified way of "like") but not "breaking off the fish's head" (specified). That is (has to be) done by "one" (an unspecified person). The touch quality inherent in "breaking off the head" is too difficult to stand for Veronique, so she is putting a distance between her and the fish, by inserting "one" for "I":

Figure 4. *Veronique on killing the codfish: Between proximity and distance.*

* **The activity of dissecting** had a periphery of adjoining activities, manifesting themselves in the videotapes recorded during the lesson and in the interview tapes and their transcripts. These activities focused, as objects, animals, and as subjects, the pupil (in an actual or virtual situation). They were modalised in a certain way. The main modality of objects appearing in the course of the interviews, and the relations to them, was *disgust* and *fright*. This was called by the pupils by name. An affective modality akin to these was also noticeable in interjections and half-expressions like "it was so...", "this way..." etc.

* *Acting persons* See figures 1 and 4.

Piotr Szybek

Discussion

The pattern emerging from viewing both levels of the story of the event is:

There is a complex of situations, actual and potential, where animals can be touched, where touching is connected to animals being killed, skinned, cut up etc. The feelings of fright, disgust etc. which seem to impede the touching of animals can be overcome by a "nice arrangement", being another way of touching, in which objects become transformed into objects not appearing in affective modalities of disgust, fear etc.

A new stage is constructed, where squid are not slimy and disgusting, but neat drawings in textbooks, and parts of squid lie on trays, arranged in a pattern corresponding to the neat drawing. This stage is constructed in textbooks and classroom activities, two manifestations of science education in the school system. (There is here an interaction of two stages. It is more like in a film, where the scene is moved to another setting).

The touching of animals is touching by humans, and humans being touched themselves. This can lead to (a) sharing of touch, and (b) delegating touch.

(a) sharing of touch, e.g. Bert waving the part of the squid he calls "testicles" at Beth.
In the interview material there are virtual touches:
- Veronique will not touch the rabbit her virtual husband has shot, but runs away.
- The virtual touch of the codfish ("breaking off its head") repulses her.
- Pauline will not cut up the "half pig" nor pluck the dead chicken.

This permits us to see the waving as a virtual touch. The character of this touch depends on whether Bert's calling the squid part "testicles" was a lapse, or not (he could have meant "tentacles").

A possibility of seeing dissection as sharing of touch emerges here. The teacher, the authors of the textbook, the pupils, are sharing the touch of a slimy, disgusting fish. They are also sharing the touch of "nice arrangement", effecting a construction of a special stage.

(b) delegating touch. Examples of this are:
- Veronique's father and uncle being made to "break off the head" of the fish,

- her virtual husband being made to skin the rabbit,
- Pauline's parents being made to skin and cut up the half pig.

All this having been done, the girls can process the "nicely arranged" animals.

The new stage constructed in science lessons shares the quality of "nice arrangement" with the kitchen, where the events spoken of in the interviews have ("virtually") occurred. There is nothing left reminding of the animal which has been hunted down and shot, or grabbed and slaughtered. The traces of fur and bristle having been removed, the only object appearing there are "nicely arranged" pieces of meat, ready for the pan. A virtual remnant of the concrete living animal is there, though, since Pauline admits an effort "not to think" without which she is unable to eat the chops from the half pig. This means that the animal body is reduced to the status of a source of unpleasant experiences, and not belonging to a being which could have some rights.

Educational conclusions

The most important quality of a stage is the *legitimacy of appearing*. On the stage where education (within schooling) is staged something may and can, something cannot and something may not occur. Here, it transpires that just as it is legitimate to eat meat, it is legitimate to dissect squid. It would even be legitimate (for some pupils, at least) to cut up a cat or a dog, if it was "as a lesson, like". This legitimacy is brought about by removing from the stage objects which might appear in affective modalities – just like the bits of fur and skin are removed after a rabbit is skinned. The "nice arrangement" is making object appearing in affective modalities not expected on the stage, and thus seeing animals in the affective modality (to pity them, or to see dissection as disgusting) becomes not legitimate. The occurrence of a "poor squid" is an event which cannot and may not be expected here.

The construction of both these stages entails setting the parameters for gender, expectations as to "boys" and "girls" (the stages interact here). It is a boy who is touching a girl with a squid. It is a boy who claims he would dissect a cat and a dog. The answers to the questionnaire showed a significant difference: most girls say they would not dissect a cat, most boys say they would. Probably the boys would "chicken out" if there was a real dead cat on the desk. The fact is rather that there is a boy/girl-talk, where boys say "cool things" like "I'd cut up a cat" and girls flinch at that. The notion of the girl/boy-talk is further supported by a significant difference in the answer to the question "Did you like dissecting the squid". Girls did not. Boys did. In connection to this, I would like to submit that there is a common talk, constituted by the interaction of the two gender components (in that school, of course, it can only be approapriate to conjecture that it might be the case in other schools, too – the practitioner can, however, become aware of the possibility). This can be described by the girls posing as "sissies" and the boys as macho, the roles being complementary. One could see Bert's waving the "testicles" at Beth in this light: Beth refuses to play "sissy" and just shrugs, shaking off the touch, as it were, which makes Bert start running around, seeking a girl who would confirm his macho-role by playing "sissy".

Pauline's "going at it", and "getting it done" seems to be characteristic of most girl pupils in that pupil group; these (a) comply with teacher's instruction, and (b) do not like dissection (as shown by the analysis of the questionnaire results). Rather than seeing the two parts, (1) the teacher and the boys and (2) the girls, as active and passive respectively, the metaphor of a tennis or soccer match can be proposed, the teacher and the boys playing "in the offensive", and the girls "in the defensive". The girls shrugging and complying makes it possible to play the game over and over.

"Nice arrangement" and "touch" occurring in this episode of science educuation are events in the phenomenological sense, as defined by Hannah Arendt, constituting a breach of routine, and a new outlook. Education can be seen as propelled by events thus understood. What we, following Piaget, have been calling "cognitive conflict", or what is broadly referred to as conceptual change (Carey 1985; Posner & Strike 1992; and Vosniadou & Brewer 1992, to give some examples) can also be seen as events, in Arendt's sense. The phenomenological view makes it possible to see the corporeal aspect of this, an aspect always concretely manifested, in concrete corporeally anchored experiences, and discernible in the affective modalities in which objects, pointed out by speech and action, appear.

Analysing events constituting a practice leads to changes in that practice. In this case, at the root of the whole enterprise was the question of what dissection of squid can contribute within the context of learning biology – and a suspicion that there is more to the course of action pursued by the pupils than learning biological subject matter. As should have become obvious in this paper, phenomenological inquiry cannot reveal hidden thoughts. The focus is rather on bringing to light potentials being constituted in the course of events. Choosing to let something happen or not amounts to the construction of a stage of events where some events become expectable and some not.

The results of this study might help other teachers look for and identify "boys' and girls' talk" similar to the one described here, in this or similar contexts. The question of course arises how desirable it might be to comply with pupil demands of this kind. It is an unnegotiable assumption that the primary and overriding objective of education has to be reproduction of a democratic society. This implies equity, not least among sexes/genders. Another question is how expectable the stage constructed in dissection makes events like respect shown to nature, and a caring attitude to the environment (two of the overriding goals stated by the Swedish national curriculum; the attitude expressed by this does not, however, seem to be something unprecedented, unique and characteristic for Sweden).

Most important, however, is the attitude of practitioners towards goals that this paper wants to promote. This entails abandoning the turn from passively implementing of a school tradition, manifesting itself in what the pupils find expectable to say, to questioning the goals implied in the course of action prescribed by that tradition. The practitioner might benefit from a way of looking at event resembling the one adopted in this paper: looking at the world as if it was always new to the observer. Is that not, by the way, a genuinely scientific attitude?

References:

Arendt, H. (1970) *On violence* New York: Harcourt, Brace & World

Carey, S. (1985) *Conceptual change in childhood* Cambridge, MA: The MIT Press

Cronbach, L.J. (1971) Test validation. In R.L. Thorndike (Ed.) *Educational measurement* (pp. 407-507) Washington, DC: American Council of Education

Cronbach, L.J. Construct validation after thirty years. (1989) In Linn, R.L. (Ed); *Intelligence: Measurement, theory, and public policy: Proceedings of a symposium in honor of Lloyd G. Humphreys.* (147-171). Champaign, IL: University of Illinois Press

Giorgi, A. (1975) An application of phenomenological method in psychology. In A.Giorgi, C.T.Fisher & E.L.Murray (Eds.) *Duquesne studies in phenomenological psychology* Vol II, 82-103, Pittsburgh: Duquesne University Press

Husserl, E. (1962) *Ideas: general introduction to pure phenomenology* New York: Collier

Piotr Szybek

Husserl, E. (1976) *Experience and Judgement*. Evanston: Northwestern University Press

Husserl, E. (1977) *Phenomenological psychology*, The Hague: Martinus Nijhoff

Husserl, E. (1980) *Collected works Vol. 1 Phenomenology and the foundations of the sciences. Book 3 Ideas pertaining to a pure phenomenology and to a phenomenological philosophy* The Hague: Martinus Nijhoff

Husserl, E. (1989) *Ideas pertaining to a pure phenomenology and to a phenomenological philosophy: First book. General introduction to a pure phenomenology* The Hague: Martinus Nijhoff

Karlsson, G. (1993) *Psychological qualitative research from a phenomenological perspective* Stockholm: Almqvist & Wiksell International

Meyer-Drawe, K. (1984) *Leiblichkeit und Sozialität. Phänomenologische Beiträge zu einer pädagogischen Theorie der Inter-Subjektivität* München: Wilhelm Fink Verlag

Moustakas, C. (1994) *Phenomenological research methods* Thousand Oaks: Sage

Ricoeur, P. (1992) Phenomenology and hermeneutics. In J.B.Thompson (ed.) *Hermeneutics and the human sciences: essays on language, action and interpretation* Cambridge: Cambridge University Press

Sages, R., Falk, K. & Johansson, C.R. (1999) *The meaning of work. A phenomenological study* Department of Psychology, Lund University, Work Science Division Monograph Series nr 2 (in press).

Sages, R. & Szybek, P. (in press) A phenomenological study of pupils' knowledge of biology in a Swedish comprehensive school. I. Presentation of the method. *Journal of Phenomenological Psychology*

Schutz, A. & Luckmann, T. (1977) *The structures of the life-world*. London: Heineman

Strike, K.A. & Posner, G.J. (1992) A revisionist theory of conceptual change. In R.A. Duschl & R.J. Hamilton (Eds.) *Philosophy of science, cognitive psychology, and educational theory and practice. 147-176,* Albany, NY: SUNY Press

Vosniadou, S. & Brewer, W.F. (1992) Mental models of earth: A study of conceptual change in childhood. *Cognitive Psychology* 24, 535-585

Section 2

Practical work and learning about science

Introduction

A commonly-articulated aim for practical work is to teach students something about the nature of science itself. Two main arguments are made:

- In order to understand specific products of the scientific enterprise (i.e. laws, concepts, theories, experimental methods and procedures), students need to know something of the nature of science itself (the 'science learning argument').
- In order to understand and participate in scientific and technological aspects of society, whether as a citizen, worker, or interested individual, it is necessary to know something about the nature of science (the 'scientific literacy argument').

Although it is hard to imagine what the *opposite* of either of these arguments might be, there is no simple answer as to how school science might address the nature of science, whether to promote students' learning of scientific concepts or to underpin scientific literacy. Firstly, experts do not agree what the nature of science is, let alone which aspects of the nature of science might be presented to school students through the science curriculum. This is in marked contrast to conceptual areas of the curriculum, where experts are much more likely to agree upon what would be a correct and appropriate presentation of a given idea to students. Secondly, although the nature of science is disputed, most serious scholars would portray science as multidimentional, emphasising the differences between the knowledge and practices of groups such as research astronomers, research and development chemists and hospital haemotologists.

This raises serious problems for those who claim that school science should be more *authentic*. The most typical portrayal of authentic science in the science education literature involves groups of individuals working to generate new knowledge or solve novel problems. But why is this *authentic*? It is open to question whether many groups of scientists, such as those working in diagnostic laboratories, engage in this kind of work. In addition, professional research scientists attend lectures and presentations by experts whose main purpose is to explain concepts clearly. So a case could also be made that lectures and practical demonstrations by teachers in school are authentic. A further question is whether practical work is the best way of teaching students about the nature of science.

These issues are raised and explored in the four papers in this section. Ntombela discusses philosophical issues about the nature of science in the context of the use of practical work in a South African science education project. The paper focuses broadly upon the scientific literacy argument. In his paper, he encounters the tension between a view of the 'authentic' practice of science on the one hand, and typical school science practical work on the other. 'Authenticity' is defined by Ntombela more in terms of very naïve views about science on the part of students and teachers, than a clearly articulated view of what *ought* to be taught about science through practical work, and the curriculum more generally.

My own paper draws upon findings from two survey-based studies and one case study to illustrate some of the differences in meaning that are likely to arise when teachers and students talk about practical work. The paper focuses on the science learning argument, illustrating the ways in which students in upper primary, secondary and university science education appear to misinterpret the nature and purpose of practical work.

Both the Ntombela and Leach papers draw mainly upon survey methods, with the obvious possibility that students' responses to questions do not predict their actions in learning settings. The other two papers in this section present case studies of teaching and learning about science. Watson, Swain and McRobbie explore how the teaching style of two teachers influences whether students take decisions during an open-ended practical activity typical of those used in UK science classrooms. Drawing upon a view of science learning as a process of enculturation into a community of practice, Molyneux-Hodgson, Sutherland and Butterfield explore what students are learning during practical activities in a work-based, vocational context. Questions are raised about the feasibility of presenting typical syllabus content in a workplace setting, and the implications of this for the role of the teacher.

In both of these papers, in common with other studies of teaching about the nature of science, it is hard to tell what learning aims the teachers had in mind for the practical activities presented. This makes it rather difficult to judge the value of the practical work in promoting student learning. By contrast, papers in the literature which address the teaching and learning of conceptual subject matter are often much more precise in identifying the learning aims underpinning teaching activities (see, for example, Tiberghien in this book). This emphasises the importance of clarity about the aims and purposes of practical work intended to teach about the nature of science.

John Leach

A marriage of inconvenience?
School science practical work and the nature of science

G. M. Ntombela

The resurgence of school science practical work that took place in the 1960s and 1970s in Western Europe and North America, and a similar trend that followed in the 1970s and 1980s in many developing countries, was seen as a way of exposing pupils to the scientific method, that is, to the 'processes of science' as used by scientists in their work.

This paper seeks to reflect on the experience of the Science Education Project, a South African Non-Government Organisation established in 1976 with the aim of transforming science education from a teacher-dominated 'chalk and talk' method of teaching, to a learner-centred, practical approach. The practical approach, while arguably having many educationally positive attributes, often fails to enhance pupils' understanding of the nature of science. Some of the reasons for this lack of success are the teachers' own shaky understanding of this aspect as well as the rigidity of curriculum materials that purport to introduce pupils to a hands-on approach.

The popular Chinese proverb: "I hear and I forget, I see and I remember, I do and I understand," may be true with regards to the understanding of certain concepts and procedures but not necessarily so in as far as the understanding of the nature of science is concerned.

School science practical work received a boost in the 1960s and 1970s when process-driven curricula became very popular. The often-quoted Chinese proverb: "I hear and I forget, I see and I remember, I do and I understand" made this passion appear quite logical. For pupils to do science and to discover things for themselves seemed to be the most appropriate way to learn science. Rather than an innovation in a true sense, this approach was more of a resurgence of the heuristic method propagated by Armstrong as early as

A marriage of inconvenience?

1903 when he insisted that "the beginner not only may but must be put absolutely in the position of an original discoverer" (quoted in Jenkins, 1979). The language used in the 1960s and 1970s also emphasised the need for a child to experience being a scientist; this s/he would do by engaging in the scientific processes to 'discover' scientific facts (or even theories) on his/her own.

In this discussion a reflective stance with regard to practical work is taken in the context of the work of the Science Education Project (SEP); a South African non-government agency that started its operations in 1976. It has, as its broad aim, the transformation of science education in Junior Secondary Schools (catering for 12-14 year olds) from a teacher-dominated, 'chalk and talk' method of teaching to a child-centred approach. The author was employed by SEP between 1985 and 1995. At the height of its teacher in-service work, this organisation reached more than 4000 teachers mainly from schools disadvantaged by the then government's apartheid policy. The strategies adopted by SEP in fulfilling its aims include the following:

- the design and provision of easy to manage, purpose-built kits that permit the performance of most school science experiments in a laboratory or ordinary classroom;
- the development of worksheets (and other curriculum materials) for use with the science kits;
- the running of in-service workshops to introduce teachers to the above materials and to help them acquire knowledge and skills essential in the practical teaching of science;
- classroom visits by SEP staff to consult and help teachers with science teaching problems.

Since 1994, SEP, like many other non-government organisations in South Africa, has drastically scaled down its operations due to unfavourable changes in the funding environment.

SEP's Philosophy and Practice

Although much has been written about the work of SEP (MacDonald and Rogan 1988, 1990; Rogan and MacDonald, 1985), scant attention has been paid to the philosophical underpinnings of its practical approach to the teaching and learning of science. The writer is not aware of any documentation that explicitly sets out SEP's philosophy of science, but what is suggested in some of the curriculum materials may illuminate the issue to a certain extent.

In the Standard 7 Physical Science worksheet book (1986, 4th edition) it is acknowledged that one of the works that influenced SEP is the American programme: *Science: A Process Approach*. Earlier SEP guidelines for teachers articulated the aims of the Project as follows:

> (...) to provide pupils with the opportunity of getting to grips with the process of science, and not serving up bits of factual knowledge about science. The aim is that the pupils should <u>do</u> and experience science, and not just hear about it. (SEP Biology Worksheets: Teachers' Edition, Pilot Edition 1980, p 1)

The same document lists the following as the processes of science: *observing; measuring; making comparisons; identifying a substance; classifying; formulating a relationship; predicting; hypothesising; verifying/designing an experiment; communicating scientific information; making models (scientifically)*.

The Teachers' Edition (1980) further asserts: "It is the mastery of these processes which enables a scientist to do science" (p 1).

The thrust of the hands-on, process approach is to shift the emphasis from 'products' of science to the 'processes' and, ideally, to introduce learners to an understanding of how science progresses (Carey and Stauss, 1970; Munby, 1982). Whilst it is conceded that the content-led approach, with its overemphasis on inert knowledge, difficult abstractions and factual recall has failed, Wellington (1988) argues that a swing to an overemphasis on processes and skills is not the solution.

The major claims made in support of the process approach include the following:

- it enhances the learning of science concepts and practical skills;
- it increases the enjoyment of science for pupils;
- it enables pupils to work like scientists and actually discover theories.

It is mainly the third claim that this discussion takes issue with because it is the one that purports to introduce learners to the nature of science.

Woolnough (1983) argues that practical work should be largely 'decoupled' from theory because when pupils are expected to discover theory from practical work "the practical is so structured that the pupil will rediscover the right answer" (p 61). It is precisely this restructuring which lessens the experience of being a scientist.

According to Woolnough (op cit.) there are three fundamental aims of practical work and they can be satisfied by three types of practicals, namely:
1. *Exercises*: used to develop certain specific practical skills like measurement and the correct general usage of the apparatus.
2. *Investigations*: used to develop scientific ways of working, using experience and understanding to solve whole investigations. Investigations should give pupils an opportunity "to take a real problem, to analyse that problem, to consider ways of attacking it, to choose the 'best' way, to implement that course of action and then to analyse and evaluate the results" (Woolnough, op cit., p 62).
3. *Experiences*: used to obtain a feel for various phenomena – important in the meaningful understanding of theory but not carried out solely for that purpose.

Although SEP worksheets tend to burden practical work with theory, pupils get ample opportunities for *exercises* and *experiences*. Open-ended investigations are, however, hard to find in these materials. It follows, therefore, that if it is the investigations which embody scientific ways of working, practical work that does not focus on them cannot claim to enhance understanding of the nature of science.

Tamir (1991) illustrates how the degree of openness and the demand for inquiry skills can be used to categorise laboratory investigations into four levels as shown in Table 1.

Level	Problem	Procedure	Conclusion
0	Given	Given	Given
1	Given	Given	Open
2	Given	Open	Open
3	Open	Open	Open

Table 1: *Levels of Inquiry in the Science Laboratory*

Most SEP worksheets (see Worksheet example) are at level 0 or 1, which confirms their categorisation as exercises rather than real investigations.

G. M. Ntombela

Worksheet example

A way of knowing there is an electric current in a circuit

Problem If you have no bulb to light up, how can you find out whether there is a current* in a circuit?

• Apparatus •

Science kit	Electricity kit	Main table
Baseboard	Cell holder and bar	4 torch cells
	Switch	
	3 connector wires	
	Nichrome wire	

Nichrome wire has a high resistance*. In other words, the electricity uses up a lot of energy to flow through nichrome wire.

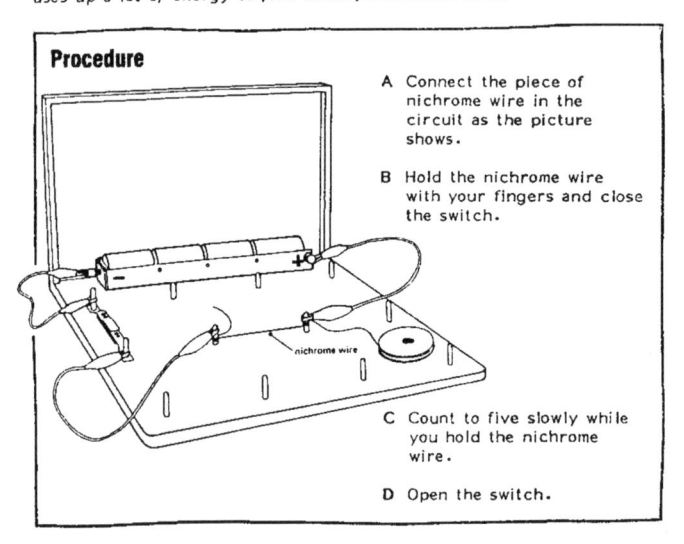

Procedure

A Connect the piece of nichrome wire in the circuit as the picture shows.

B Hold the nichrome wire with your fingers and close the switch.

C Count to five slowly while you hold the nichrome wire.

D Open the switch.

Questions

1 How did the nichrome wire feel **before** you closed the switch?

2 How did the nichrome wire feel **after** you closed the switch?

3 What do you think caused the change in the nichrome wire?

4 Where did the heat energy come from?

5 When a bulb lights up we know that there is a current in the circuit. What is another way to find out whether there is a current in a circuit?

A marriage of inconvenience?

In fairness to the SEP curriculum-materials developers it must be mentioned that there has always been a recognition that the worksheets should eventually become more open-ended and move up to at least level 2 on the above levels. The Teachers' Edition of the SEP Biology Worksheets (1980: p 26) states: "The ideal situation would be to pose the problem only, and leave it to the pupils' own initiative how to solve it."

Millar (1991) proposes a differentiation of 'practical skills' into three sub-categories as shown in Table 2.

'PRACTICAL SKILLS'

General Cognitive Processes	Practical *Techniques*	Inquiry *Tactics*
observe; classify; hypothesise etc.	measure temperature to within 1°C; separate solid and liquid by filtration etc.	repeat measurements; draw graph to see trend in data; identify variables to alter, measure, control...

◄──── (cannot be taught) ────► ◄──── (can be taught and improved) ────►

Table 2: *Sub-Categories of 'Practical Skills'*

This way of looking at practical work highlights what Todhunter warned about in 1873, namely, that: "Little of what is characteristically valuable in experimental philosophy is susceptible of transmission" (cited in Layton, 1990: p 48). The cognitive processes (observing, classifying etc.) have their *raison d'être* in whole investigations, but have limited value as ends in themselves, and cannot be taught on their own (Millar, 1989; Woolnough, 1989).

In general, SEP worksheets concentrate on 'practical techniques': for example, how to use an ammeter, how to measure resistance, how a variable resistor works, how a manometer works etc. 'Inquiry tactics' found in the worksheets have more to do with the transformation of data like tabulating and graphing than with elements of a good inquiry such as the control of variables.

The role of the teacher

In the implementation of any curriculum innovation, the primacy of the teacher's role cannot be overemphasised. Herron (1971) maintains that the teacher acts as "ultimate interpreter" of curriculum materials since she/he is the "final filter" through which these materials pass from producer to consumer. Even with specified details of how the innovation should operate, teachers are known to personalise materials and plan their own classroom routines (Brown and McIntyre, 1982). It is this reality that leads Baird (1988: p 70) to assert: "The future of science education does not lie primarily in curricula or in technology. It lies in the teacher of science."

Stenhouse (1985) expresses the view that

> (...) the widely held, often unstated and to some extent unconscious assumptions about science and science education, whatever these assumptions may be, exact a strong influence upon the way we educate, the way we do science, and the way we research into science education. (p. 10)

The pivotal role of the teacher was already recognised in 1900 when heurism was under attack and "there were unmistakable signs that, in the hands of inexperienced or inadequate teachers, the heuristic method was being misapplied," and the teachers were being seen as "the method's weakest point" (Jenkins, 1979: p 43).

A survey conducted by the author in 1993 revealed that the view of science held by most science teachers implementing the SEP curriculum and student-teachers at a College of Education in South Africa, are, in many respects, at odds with current thinking in the philosophy of science (Ntombela, 1993).

The common view held by these teachers and student-teachers concerning scientific theories is that:

- they represent the truth;
- they are developed from observed data;
- they are formulated by scientists who may or may not be objective;
- they do change over time.

There is also a strong belief that by following steps given in worksheets pupils can 'discover' the theory for themselves.

Science syllabi in South Africa prior to 1994 invariably specified that pupils should be helped to develop the ability to observe objectively and to solve problems by applying a scientific method of reasoning and scientific procedures. When teachers and student teachers were asked to unpack this lofty aim, it was obvious from the responses or lack of responses that they had seldom reflected on these ideas.

Many process-oriented guidelines start with 'observation' when listing scientific processes and this tends to give the impression that 'observation' is the starting point in any scientific endeavour. In line with this perception of the role of observation, many teachers expect pupils to "observe what is put in front of them before the teacher tells them what it is all about" or to observe "things which they have not been taught about" (Ntombela, op. cit.). A common confounding experience for teachers therefore takes place when pupils, during practical work, do not observe what they are expected to.

Scientific theories and observations

Kosso (1992) gives a simple yet profound description of theory when he explains that it is "an answer to a 'what's going on here?' question" and "an account of the underlying composition and the unseen causes of the world as we experience it" (p 15). Theories, he continues, "have in common the feature of speaking about things beyond evidence. They take risks in telling us more than we can see for ourselves."

Theory influences observation. In experimentation:

> The business of gathering data is neither comprehensive nor random, so it must be directed by theories that tell us where to look and what is important enough to warrant attention. (Kosso, 1992: p 114)

Information that is gathered from observation is not simply given by nature but rather actively taken by the observer who is guided by theories, beliefs and concepts that shape these. Koyre (1968) argues:

> a mere collection of observational or experiential data does not constitute a science: they have to be ordered, interpreted, explained. In other words, it is only when subjected to theoretical treatment that a knowledge of facts becomes science ... the empiricism of modern science is not experiential; it is experimental... Experimentalism is a teleological process of which the goal is determined by theory. (p. 89)

The expectation that pupils can use observation-led processes to rediscover scientific theories misrepresents the nature of science. The criterion that is generally used in classroom experiments is that they should work; and this is sensible if one considers the number of experiments or 'discoveries' that pupils have to go through in a term or year. It is this very criterion that weakens the understanding of how 'real' science happens. Everyone participating in the exercise becomes preoccupied with the 'right' theoretical knowledge. 'Discovery learning' cannot, therefore, offer a multi-purpose pedagogical solution to the need to know scientific theories as well as the need to understand the processes by which science grows (Layton, 1990).

Nercessian (1991) argues that even the term 'discovery' in the context of scientific experiments is also philosophically misleading because the process is better characterised as 'invention' because "scientific representations are constructed – they are made, not stumbled upon or found" (p 145). This view is shared by Ravetz (cited in Jenkins 1979), who maintains that the logical and imaginative operations in science are conducted with "intellectually constructed things and events" and not with objects of common-sense experience.

What is generally known as the *under-determination of theory by observation* also makes it difficult to see how pupils' observations can lead them to the discovery of theories. This is the idea that "more than one theoretical construction can always be placed upon a collection of data" (Gjertsen, 1989).

Theory and Truth

In the 'discovery learning' type of practical work the notion of truthfulness of scientific knowledge is often reinforced by the pre-planned design to arrive at the 'right' answer. The expectation that everybody will get the same answer further proves that it is the only 'right' solution to the problem (Carter, 1989).

The concept of 'truth' when describing scientific theories is perhaps not a fitting one. As Munby (1982) points out: "theories and explanations are not, strictly speaking, true or false, right or wrong; rather they conform or do not conform with our observations, and they are judged on their usefulness to our attempts to construct a predictable, scientific world about which we can make generalisations"(p 29). Theories, therefore, become disused when they become less useful than when they were developed. Striving for increasingly useful constructions is not the same as striving for absolute truth. The refutation of an already established theory usually amounts to the discovery of limits to its applicability (Settle, 1991). Chalmers (1990) argues that: "modern science has replaced the utopian aim for certainty by the requirement for continual improvement or growth"(p 36).

An educational response

A meaningful approach to school science practical work should reflect our understanding of the nature of the subject (science), utilise what we know about teaching and learning and take cognisance of the contextual realities. Having argued that the 'process approach' portrays a somewhat confused nature of science one needs to heed the warning that:

> we would do children a gross disservice if we implied that the world of the scientist is totally anarchic – a disservice just as great as the suggestion that science is propelled by a single all-powerful method. Science does have methods, but the precise nature of those methods depends on particular circumstances. (Hodson, 1991: p 24).

Investigations

If pupils are to learn scientific ways of working, they have to tackle whole investigations, where a premium is not placed on the correct, textbook answer (see earlier arguments advanced by Woolnough, 1983). They (the pupils) have to tackle problems that are of interest to them. It should not be a pre-requisite that such problems can readily be linked to a particular syllabus topic because this will be tantamount to overburdening the investigation with theory. In solving such open-ended investigations, pupils will not focus on a standardised method or procedure but will begin to appreciate the role of diverse approaches, imagination, planning, confirmation, and instrumentation in the pursuit of scientific knowledge (Matthews, 1988).

Pupils should be given a chance to appreciate that the methods of inquiry depend on the nature of the problem being investigated, and what is learned (the knowledge gained) also depends on the methods used. The document *Science for All Americans* (1990) concludes:

> Science teaching that attempts solely to impart to students the accumulated knowledge of a field leads to very little understanding and certainly not to the development of intellectual independence and facility. But then to teach scientific reasoning as a set of procedures separate from any particular substance – 'the scientific method', for instance – is equally futile. (p. 203)

History and philosophy of science (HPS)

The use of the HPS in science teaching can help "paint a richer, and more variegated, picture of science than has been portrayed as standard in school texts and classrooms" (Matthews, 1990: p 45). Educators and learners can be enabled to better appreciate the role of theory, observation, experiment and models in the advancement of science. Documented accounts of the great works of people like Galileo, Darwin, Millikan, Huygens, Flemming, Mendel, Dalton and many more could be powerfully educative; especially because many of the great scientific works demonstrate "the apparent indifference of scientific theory to unfavourable evidence" (Gjertsen, 1989: p 245).

Through HPS, educators and learners can be reminded that "scientists often bring with them to their work, along with a variety of cultural and intellectual presuppositions, excessive amounts of personal ambition," and that what are supposed to be objective judgements are often influenced by many extraneous factors (Gjertsen, 1989).

HPS can facilitate the understanding that "scientific ideas and theories not only result from the interaction of individuals with phenomena but also pass through major social institutions of science before being validated by the scientific community" (Driver, 1989: p 85).

Arguing in support of the inclusion of HPS in school science, Elkana (1970) asserts:

> (...)history of science should become an integral part of science teaching at all levels, not instead of the systematic exposition of the science but in addition to it. It is through history of science that students can become aware of the open nature of science, and more importantly of the open nature of methods by which one can do science. (p. 35)

Science is an on-going human endeavour and many present-day scientific events offer deep insights into its nature. In South Africa, the claims around the Virodene cure for AIDS, how these were censored by the scientific community (Medical Research Council) as well as the support of the Virodene researchers by the government's Health Department provide a good account of how science happens.

The curriculum

For many years the science curriculum in South Africa has been strong on content and paid lip-service to skills and processes. Assessment, by focusing on factual knowledge, also served to reinforce this approach. Since 1994, South Africa has a radically different curriculum known as OBE (Outcomes-Based Education). This approach represents a shift from focusing on teacher/trainer inputs aimed at helping learners master syllabus content, to a holistic approach addressing knowledge, skills and processes. Olivier (1997) explains the balance as follows:

> One of the important components of the learning process will always be to master knowledge and skills as well as being assessed on it. Neglecting this component will place more emphasis on processes. This will furnish learners mainly with methods to follow, without being able to attach substance and body. Focusing only on processes will deprive learners of the capacity to construct and achieve outcomes in other circumstances as will an approach focusing mainly on mastering content. (p. 30)

In the OBE curriculum the 'Earth and natural science' course does not specifically set out the content but outlines four broad themes: the planet Earth and beyond; life and living; energy and change; and matter and materials. It represents a shift towards "the development of an understanding of scientific concepts that will enable learners to use science, to critically reflect on it, and to see how it relates to their own lives in general" (Maurice-Mopp, cited in Gledhill, 1998: p 4).

Teacher Education

The high ideals of a curriculum come to nothing if the teaching corps does not have the capacity or the confidence to implement them. This problem is exacerbated if the curriculum itself has serious flaws, as is the case with 'discovery learning' approaches. It is the teachers' views of the nature of science, implicit or explicit, rather than the official view of the curriculum that is reflected in the teaching methods employed (Selley, 1986).

A successful science teacher, according to a 1929 text (cited in Matthews, 1990) is someone who:

> knows his own subject ...is widely read in other branches of science ... knows how to teach ... is able to express himself lucidly ... is skilful in manipulation ... is resourceful both at the demonstration table and in the laboratory ... is a logician ... is something of a philosopher ... is so far an historian that he can sit down with a crowd of boys and talk to them about the personal equations, the lives, and the work of such geniuses as Galileo, Newton, Faraday and Darwin. (p. 40)

The Centre for the Advancement of Science and Mathematics Education (CASME) is a university-affiliated organisation that offers formal/certificated courses to teachers of secondary school science and mathematics. In an effort to produce teachers of the type described in the document cited by Matthews above, modules dealing with the history and the nature of these subjects are taught. The aim of such modules is to equip teachers, not only with an understanding of the nature of science but also with an ability to share that understanding with learners in a meaningful and experiential way. The majority of teachers are not exposed to these areas of study during their pre-service education, hence the views expressed in the survey quoted earlier (Ntombela, 1993).

Conclusion

If we believe that scientific literacy is necessary for a country's responsible and sustainable development, then a continuous review of science education in schools and other settings is needed. In almost all countries, developing and developed, "science education without some laboratory experience is unthinkable" (Layton, 1990). The resources poured into the provision of these learning facilities, will not have been effectively utilised if they are used in a manner that not only fails to enhance the learners' understanding of the natureof science but actually distorts it. All science learners should get a good grasp of how the scientific enterprise operates, and have an appreciation of its achievements and its limitations.

References

American Association for the Advancement of Science. (1990). *Science for all Americans*. Washington D.C.

Baird, J. R. (1988). Teachers in Science Education. In Fensham, P.J. (Ed), *Development and Dilemmas in Science Education*. East Sussex: The Falmer Press, pp 55-72.

Brown, S. and McIntyre, D. (1982). Costs and Rewards of innovation: taking account of teachers' viewpoint. In Olson, J. (Ed), *Innovation in the Science Curriculum*. Croom Helm Curriculum Policy and Research Series, pp 107-139.

Carey, R. L. and Stauss, N. G. 1970). An analysis of experienced science teachers' understanding of the nature of science. *School Science and Mathematics*, 70, pp 366-376.

Carter, C. (1989). Scientific knowledge, school science, and socialization into science: issues in teacher education. In Hergert, D.E. (Ed), *The History and Philosophy of Science in Science Teaching: Issues in Teacher Education*. Proceedings of the International Conference, Florida State University.

Chalmers, A. (1990). *Science And Its Fabrications*. Milton Keynes: Open University Press.

Driver, R. (1989). The construction of scientific knowledge in school classrooms. In Millar, R. (Ed), *Doing Science: Images of Science in Science Education*. London: The Falmer Press, Chapter 4: pp 83-106.

Elkana, Y. (1970). Science, philosophy of science and science teaching, *Educational Philosophy and Theory*, 2, pp 15-35.

Gjertsen, D. (1989). *Science and Philosophy: Past and Present*. London: Penguin Books.

Gledhill, L. (1998). Science gets a new life in schools. *Mail And Guardian Supplement*, February 27 to March 5, p 4.

Herron, M. D. (1971). The nature of scientific enquiry. *School Review*, 79, pp 171-212.

Hodson, D. (1991). Philosophy of Science and Science Education. In Matthews, M. R. (Ed), *History, Philosophy, and Science Teaching: Selected Readings*. Toronto: Oise Press, pp 19-32.

Jenkins, E. W. (1979). *From Armstrong to Nuffield*. London: John Murray.

Kosso, P. (1992). *Reading the Book of Nature: An Introduction to the Philosophy of Science*. New York: Cambridge University Press.

Koyre, A. (1968) *Metaphysics and Measurement: Essays in Scientific Revolution*. London: Chapman and Hall.

Layton, D. (1990). Student laboratory practice and the history and philosophy of science. In Hergaty-Hazel, E. (Ed), *The Student Laboratory and the Science Curriculum*. London: Routledge, pp 37-59.

MacDonald, M. A. and Rogan, J. M. (1988). Innovation in South African Science Education (Part 1): Science teaching observed. *Science Education*, 72 (2), pp 225-236.

MacDonald, M. A. and Rogan, J. M. (1990) Innovation in South African Science Education (Part 2): Factors influencing the introduction of instructional change. *Science Education*, 74 (1), pp 119-132.

Matthews, M. R. (1988). A role for history and philosophy in science teaching. *Educational Philosophy and Theory*, (20) 2, pp 67-79.

Matthews, R. (1990). History, Philosophy and Science Teaching: a rapproachment. *Studies in Science Education*, 18, pp 25-51.

Millar, R. (1989). What is 'scientific method' and can it be taught? In Wellington, J. J. (Ed), *Skills and Processes in Science Education: a critical analysis*. London: Routledge, pp 47-62.

Millar, R. (1991) A means to an end: the role of processes in science education. In Woolnough, B. E. (Ed), *Practical Science*. Milton Keynes: Open University Press, pp 1-7.

Munby, H. (1982). *What is Scientific Thinking?* Guidance Centre: University of Toronto.

Nercessian, N. J. (1991). Conceptual change in Science and in Science Education. In Matthews, M. R. (Ed), *History, Philosophy, and Science Teaching: Selected Readings*. Toronto: Oise Press, pp 133-148.

Ntombela, M. (1993). The Case For Philosophically Grounded School Science Practical Work With Special Reference to the Science Education Project. Unpublished M.Ed. Thesis: Leeds University.

Olivier, C. (1997). *Outcomes-Based Education And Training Programmes*. Ifafi: OBET Pro.

Rogan, J. M. and MacDonald, M. A. (1985). The In-Service Teacher Education Component of an Innovation: A case study in an African setting. *Journal of Curriculum Studies*, 17 (1), pp 63-85.

Selley, N. J. (1986). The place of alternative models in School science. In Brown, J., Cooper, A., Horton, T., Toates, F. and Zeldin, D. (Eds), *Science in Schools*. Milton Keynes: Open University Press, pp 121-130.

Settle, T. (1991). How to avoid implying that physicalism is true: a problem for teachers of science. In Matthews, M. R. (Ed), *History, Philosophy, and Science Teaching: Selected Readings*. Toronto: Oise Press, pp 225-234.

Stenhouse, D. (1985). *Active Philosophy in Education and Science*. London: George Allen and Unwin.

Tamir, P. (1991). Practical work in school science: an analysis of current practice. In Woolnough, B. E. (Ed), *Practical Science*. Milton Keynes: Open University Press, pp 13-20.

Wellington, J. J. (1988). The place of processes in physics education. *Physics Education*, 23, pp 150-155.

Woolnough, B. E. (1983). Exercises, investigations and experiences. *Physics Education*, 18, pp 60-63.

Woolnough, B. E. (1989). Towards a holistic view of processes in science education. In Wellington, J. J. (Ed), *Skills and Processes in Science Education: a critical analysis*. London Routledge, pp 115-134.

Learning science in the laboratory
The importance of epistemological understanding

John Leach

When students carry out labwork they do so based on assumptions about the purposes of the task in hand, the nature of the data that might be collected and their relationship to knowledge, the nature of explanation and investigation and so on. This paper draws upon recent work which addresses the epistemological and ontological understanding that science students in upper primary and secondary schools, and universities, draw upon during practical work, and the impact that these images of science might have upon students' learning during practical work.

Epistemology and the teaching laboratory

It has become very fashionable in science education to call for better understanding of 'the nature of science' amongst students. The major journals in science education have all published research papers on the nature of science in science education in recent years, one journal is devoted entirely to links between philosophical issues in science and science education, and major conferences have been organised on the theme. In addition, curriculum priorities in some countries are placing more emphasis upon 'the nature of science' (e.g. DES, 1991; TTA, 1998, Rutherford and Ahlgren, 1989).

The literature addressing what students ought to know about the nature of science and why they ought to know it contains two major lines of thinking. The first of these addresses the science curriculum as a vehicle for promoting scientific literacy, and the types of knowledge about the nature of science that might be included in such a curriculum (e.g. Driver et al. 1996). The argument that is presented states that students need some meta-understanding of the nature of scientific content knowledge in order to be able to interact with it when encountered in the media or in professional settings, and that a general education should include something relating to science as a cultural activity.

The second line of thinking focuses more upon students' learning during science lessons. Understanding scientific content goes beyond simple recall of laws, facts, hypotheses, models and theories from a scientific discipline. Some understanding of the nature of that content is also required. This might involve recognising that thermodynamic accounts of chemical behaviour are in fact models based upon abstractions such as infinitely large energy sinks, or recognising that the point masses and friction free planes of Newtonian mechanics are abstract models. In order to use abstract models to account for the behaviour of real objects, it is necessary to consider how the point masses, friction free planes or energy sinks of the scientific world can be related to objects and events in the material world. Ontology is the branch of philosophy addressing the status of the entities used in knowledge systems, and their relationship with entities in the material world. It is therefore argued that in order to appreciate how scientific knowledge is used, students need some rudimentary understanding of its ontology.

But scientific content is only one facet of scientific understanding. Some methodological understanding is also involved, including aspects such as recognising the strategies by which enquiries are tackled within particular scientific disciplines, and knowing about the established procedures and routines used in carrying out enquiries. There is no one approach to empirical enquiry in science. Within a field, this might involve having some idea about what counts as an appropriate explanation, the sorts of data that are drawn upon in investigations, and the accepted warranting of knowledge claims. It might also involve coming to understand specific techniques used within disciplines for collecting and analysing data. Epistemology is the branch of philosophy which addresses the reasons why knowledge claims are believed to be reliable. In effect, to understand why data are collected and how they are interpreted within a scientific discipline, students have to have some rudimentary knowledge of epistemology. In addition, there is a recognition amongst contemporary philosophers, historians and sociologists of science that social and institutional processes are critically important in determining what comes to be agreed as reliable public scientific knowledge, and who comes to be accepted as an expert when opinions are sought on scientific and technical issues. Scientific understanding might therefore include some rudimentary understanding of the social and institutional mechanisms through which science operates and interacts with broader social interests.

Although there is some degree of consensus amongst science educators that scientific understanding involves some rudimentary knowledge about 'the nature of science' (i.e. its epistemology, ontology and sociology), and that learning science therefore involves learning something about the nature of science, there is a marked absence of consensus about what exactly students need to know about the nature of science and how it might be taught and learnt. Some argue for an explicit presentation of philosophical issues to students though the rationale for this is usually to enhance students' appreciation of science as a cultural product rather than to enhance their learning of science content (e.g. Matthews, 1994, Bandiera et al., 1998). However, there are acknowledged difficulties in considering how philosophical content might be introduced into the curriculum given that there is no canonically agreed body of philosophical knowledge about science.

This paper addresses the specific context of practical work used in teaching science. An argument is presented that school practical work requires students to draw upon particular ontological and epistemological representations, even though these are not typically made explicit to students. Examples are presented of some fundamental ontological and epistemological representations, which science teachers, scientists and philosophers of science would more than likely agree as appropriate to be drawn upon in particular cases. By contrast, the reasoning used by students is often based upon inappropriate ideas about ontology and epistemology, resulting in actions being taken that science teachers, scientists and philosophers would agree as inappropriate. Finally, suggestions are made as to how students' learning could be improved by addressing some of the basic ontological and epistemological demands made through routine practical work in teaching.

Identifying learning aims for students during practical work

Teaching and learning in the laboratory

There are wide differences across Europe in the ages at which practical work is first used as a teaching method for science students (Tiberghien et al., 1998). In the UK, students do some practical work in primary school, and practical work is a common teaching method from the beginning of secondary school at age 11. In other countries, it is more typical for practical work to be used from the later years of secondary school, or even from the beginning of university study. Practical work is carried out to address various learning goals, relating to learning scientific content and learning about the methods of science (Millar et al., in this book). How does teaching and learning happen during practical work in these varied settings?

I am persuaded by the view that learning has both *social* and *individual* components, following Vygotsky and Bruner. In this view, cognition originates on the *social* plane, and is then *internalised*. Formal knowledge (such as scientific knowledge) is differentiated from the spontaneous knowledge used in everyday situations. The learning of formal knowledge is portrayed as a process of enculturation. Learners are directly introduced to formal knowledge, primarily through language. This process involves more than one person and is therefore on the *social* plane. The process of internalisation involves individuals in appropriating knowledge from the *social* plane for private use. In other words, the mechanisms by which individuals learn to use scientific concepts to explain natural phenomena have to be explained at both the *social* level and the *individual* level.

In this perspective, science teaching is viewed as a process of *assisted performance* whereby experts (or more able peers, texts etc.) provide support for learners in introducing them to formal knowledge. *Scaffolding* is a special case of assisted performance. During scaffolding, teachers address their comments to the particular needs of learners in given situations, with the result that students are able to perform a given task with help. [Teaching is clearly on the *social* plane.] Internalisation happens when teachers remove assistance, and the students are still able to perform the task. [In other words, students become able to operate on the *individual* plane.]

In his re-interpretation of Vygotsky's work, Bruner (1962) suggests that teachers act as 'vicars of the culture'. That is, they are senior members of a culture, responsible for bringing others into a culture. In the case of science education, science teachers might be thought of as representatives of the culture of science, whose role is to make (some of) the practices and ideas of science accessible to students. This model of teaching suggests that 'good' teaching involves teachers in planning how to assist students so that they can understand various aspects of science, given existing levels of understanding. But Vygotsky and the neo-Vygotskians have given little attention to the nature of students' existing knowledge and how it might affect teaching and learning – either in terms of the assistance that teachers ought to give, or the factors that might make internalisation difficult to achieve. However, research now exists that provides information about some of the ways in which students are likely to reason during practical work, as described below.

Examples of students' reasoning in the laboratory

During practical work, it is assumed that students and teachers share common understandings about epistemological and ontological issues such as the purpose of investigation, the ways in which scientific models or theories are used to explain the behaviour of material objects and events, and the ways in which data are collected, analysed and used in drawing conclusions. There is now a substantial body of evidence to suggest that students do not share teachers' assumptions about such issues in the absence of direct teaching. This is illustrated with reference to two studies, the first of which focuses on students in the 9-16 age range, the second focusing on older students aged 16-20.

Leach et al. (1993) investigated how British students conceptualised the process of empirical enquiry. Pairs of students aged 9, 12 and 16 (n=30 pairs at each age) were presented with a set of 9 cards, each of which contained a brief description of an activity. The activities differed along a number of dimensions:

- The extent to which school science was emphasised in the context of the activity (e.g. following a recipe to bake a cake as opposed to following a recipe to grow a crystal)
- The extent to which investigative activities followed a planned routine (e.g. trying to make a radio work by a random process as opposed to finding out which of three paper towels can absorb most water through a controlled investigation)
- The extent to which activities were undertaken against a stated theoretical background (e.g. weighing a parcel to determine which stamp to stick on to it, measuring rainfall to test a vague idea about seasonal patterns in rainfall, investigating a formally stated hypothesis about empirical generalisations such as relationships between particle size and dissolving or type of material and conductivity, and testing a simple theory about the behaviour of gases on heating).

Learning science in the laboratory

The students had to classify each activity as an experiment or not an experiment, or alternatively to state that they were unsure whether the activity was an experiment. An interview was then conducted to determine what features of the activity the students drew upon in judging whether the activity was an experiment or not, and in describing the process of empirical enquiry. The focus of data analysis was upon the representations of empirical enquiry used by the students, rather than the meanings that they attributed to the word 'experiment'. Indeed, in British schools the word 'experiment' is commonly used to mean any practical activity undertaken in science lessons.

Three main representations of the process of empirical enquiry were noted amongst the students. At the lowest level, some students did not appear to differentiate between activities with no investigative component (such as following a recipe to make a cake or crystal), and activities with the aim of finding something out. A closely related pattern of reasoning involved students in stating that the purpose of experiments was to make a phenomenon happen, often in ingenious ways, rather than to make a phenomenon happen *in order to find something out about it*. For example, such students stated that growing a crystal or making a lamp light were both experiments because something happened as the result of following a routine procedure.

The most common representations of the process of empirical enquiry

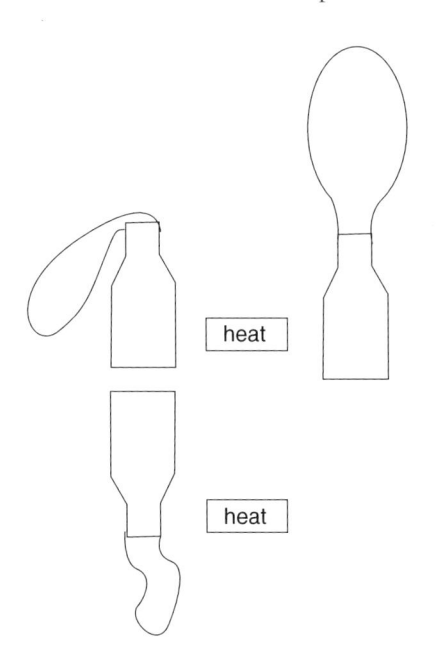

When the bottle is heated the balloon fills with air. This could be because the air *expands* when it is heated, or because *hot air rises*. This person is heating the bottle upside down to find out which idea is correct.

were based upon a recognition by students that something is *found out* during enquiries. However, the *finding out* involved was restricted to recognising simple relationships between observable variables. This is well illustrated by students' responses to one card, illustrated in figure 1, which was overwhelmingly classified as an experiment. Many students stated that this activity was an experiment on the grounds that the students were trying to find out the effect of the orientation of the bottle upon whether the balloon would inflate. No reference was made by such students to the statement about air expanding and/or rising.

For students using this form of reasoning, there was a clear differentiation between activities in which *something* is found out, and activities which are not investigative. This is in marked contrast to the students who did not appear to recognise any distinction between investigative and non-investigative activities. However, in both cases many students used the word 'experiment' to describe activities, in common with the use of the word in many UK science classrooms.

Some students indicated that they understood the presented account of the 'balloons' activity as an experiment to test between two stated explanations. In comparing students' responses on this and a number of similarly focused tasks, Driver et al. (1996) noted that 16 year old students in the sample only rarely accounted for the bahaviour of simple phenomena and events in terms of theoretical models involving posited entities such as molecules in random motion.

It was interesting to note that pairs of students in this study drew upon different representations of empirical enquiry to justify their classifications of different activities as experiments or not experiments. For the majority of students, there was no evidence that a single, strongly-held representation was being drawn upon. It is, of course, entirely appropriate to use multiple representations given the diverse kinds of empirical investigations undertaken in science. However, amongst the 16 year old sample in the study only 50% of students' justifications for classifying the balloon activity as an experiment referred to the testing of explanations. 26% of justifications stated that the activity was an experiment as the orientation of a bottle upon the inflation of the balloon was being investigated.

These findings raise serious questions about what the students who were interviewed in this survey might think that they were doing during school science practical work. During investigative practical work, it is open to question whether students would share their teachers' understanding of the purpose of the investigation. It is also doubtful whether they would recognise links between actions taken and data collected, and the scientific models and theories that provide the background to the activity.

Recent research provides evidence that significant numbers of upper secondary and undergraduate science students reason in similar ways to younger students in making links between experimental data and knowledge claims. Leach et al. (1998) surveyed 721 European science students aged 16-20 with a view to finding out, amongst other things, the ways in which the students thought of scientific knowledge claims as being related to data.

Three major representations were noted. In the first of these, the processes of measurement and data collection are viewed as simply involve 'copying' from reality, and so the process of drawing conclusions is a simple one of stating what happened in terms of data from an experiment. The need for judgements about spread, error, and so on is not recognised. This is similar to the representation identified amongst younger students who stated that investigations involve finding out whether simple observable variables are related. For example, students were presented with a data set on which measured values of electrical resistance of a superconductor were plotted at different temperatures, with error bars. Students were asked to select one of four descriptions of the best way to propose a relationship between resistance and temperature, or to state their own alternative view. The statements were:

- Draw a line joining each of the points. We are confident about each measurement, so this is the best approach
- Use a computer to generate the best curved line through the data points. This is the best approach.
- Consider which model could be used to explain this data set. Once the best model has been agreed upon, a line can then be drawn through the data points.
- There is no way of knowing which is the best way to join the data points. It is up to individual scientists to make up their own minds.

26% of upper secondary students and 28% of university students selected one of the first two statements, which contain no reference to underlying theory.

Further evidence was provided from another question that students view the relationship between measured data and the material world in terms of simple correspondence. Students were presented with two data sets each containing nine measurements of the mass of aliquots of oil. Mean values were calculated for each set. The nine values in one data set had a significantly wider spread than those in the other data set. Students were asked to state which data set most confidence could be attributed to, with a justification. 11% of upper secondary students and 6% of university students gave a response which referred to two measurements having identical values, indicating that measurements with identical values were seen as the best indicator of the 'real' value.

The second major representation noted amongst older science students was not noted amongst younger students. It involved a radical relativist view, where the process of drawing conclusions in experimental work is viewed as inherently problematic to the extent that, at the end of an experiment, it is up to every individual to believe what they want to as data cannot be used to judge any view as better or worse than any other. For example, 21% of upper secondary students and 6% of university students selected the statement: *'There is no way of knowing which is the best way to join the data points. It is up to individual scientists to make up their own minds.'* Of course, students might select this statement for a variety of reasons. Some, for example, may select it to convey the view that there is no unique way to interpret data sets. However, the fact that such a large number of students selected this statement from the four presented raises the possibility that some students do use radical relativist views in some situations, a finding reported in other studies (see Désautels and Larochelle, 1998).

The third major representation involves a view that what scientists believe, what they do during labwork, and the data that they have are all related, each one being potentially capable of influencing the other. Although this was the most commonly noted representation amongst the sample of older specialist science students, only 38% of upper secondary and 42% of university students selected the statement: *'Consider which model could be used to explain this data set. Once the best model has been agreed upon, a line can then be drawn through the data points.'*, and only 50% of upper secondary and 61% of university students stated that confidence should be attributed to data sets on the grounds of spread of the whole data set as opposed to characteristics of individual values.

Some caveats...

The evidence presented in this section drew upon examples of students' responses to survey tasks relating to laboratory situations, rather than examples of students' actions during teaching and learning situations. There is an obvious advantage, in terms of ease of data collection, in being able to present an identical survey task to large, designed populations rather than being dependent upon finding appropriate contexts for research in real teaching and learning situations. However, questions inevitably arise about the extent to which students' responses to survey questions can be used to predict their actions during practical work. The use of case study methods have shown that students exhibit reasoning and behaviour during practical tasks that is very similar to that predicted by findings from the above surveys (see, for example, Lubben and Millar, 1996, Séré et al., 1993).

Some implications for teaching science through practical work

In the above studies, it is interesting to note that students' responses are context-dependent. For example, the factors seen as critical by younger secondary students in deciding that making a cake is not an experiment are different from the factors seen as critical in deciding that the 'bottles and balloons' activity was an experiment. There is now convincing empirical evidence that individuals draw upon a number of representations of the nature of science in answering questions and performing tasks, and that there are good reasons for viewing science as a diverse and multifaceted practice (e.g. Driver et al., 1996; Leach et al., 1998; Mortimer, 1993).

In the psychological literature, considerable attention has been given in recent years to the question of whether there are developmental constraints upon students' abilities to deploy appropriate reasoning on logical tasks involving the evaluation of knowledge claims against evidence (e.g. Kuhn et al., 1988; Carey et al., 1989; Samarapungavan, 1992; Metz, 1991; Koslowski, 1996; Leach, in press). For example, Samarapungavan (1992) presents evidence that pre-adolescent children are capable of selecting theories on the grounds of empirical consistency with data, even though they may not spontaneously do so. Leach (in press) has shown that although many students in a sample of 9-16 year old students did not always take account of available evidence, or contradicted previous arguments in selecting explanations and generating predictions, all students in the sample showed themselves capable of taking full account of evidence in a logical way on at least one occasion.

There is a growing consensus that the relatively poor achievement of students (and particularly older students) in carrying out tasks involving the evaluation of knowledge claims against evidence is probably attributable to a lack of appreciation by the students as to what constitutes an appropriate strategy for the given task rather than developmental constraints on students' abilities. In practical work, the issue for teaching is often a matter of helping students to deploy appropriate forms of reasoning in given contexts through the use of more targetted pedagogy. This can be illustrated in the case of students' learning about data analysis during an undergraduate course in biochemistry (Leach, 1998). As part of an introductory laboratory course, students had to undertake a data analysis activity in which they had to process presented kinetic data from an enzyme-catalysed reaction to draw a Lineweaver-Burke plot. The values of the kinetic constants Vmax and Km were then calculated from this plot. The constants are significant in indicating the nature of binding between enzymes, substrates and inhibitors, and the rates of reactions in metabolic pathways, in terms of the Michaelis-Menten model of enzyme kinetics. The process of drawing the Lineweaver-Burke plot was presented to the students as an algorithm, and no attempt was made during the teaching to encourage students to focus upon the origins and significance of the presented kinetic data, the confidence that could be attributed to estimates of Vmax and Km, and relationships between the data handling approach and the Michaelis-Menten model.

At the end of the activity, the students' understanding of the method of data handling and the significance of the constants calcluated in terms of the Michaelis-Menten model was predictably low. From amongst a sample of 48 students, 38 appeared to have little idea as to how the presented kinetic data might have been collected. In addition, when asked how a value of the rate of a reaction should be selected from a set of values, several students referred to the significance of repeat measurements and 10 students stated that the only way to know for sure is to look in a data book! There was evidence that although many students stated that a mean value should be selected, many were following a routine algorithm that they had been taught with no understanding about why the mean was selected. Indeed, during interviews many students reverted to comparing individual measured values to justify their responses.

It is hardly surprising that exposing students to data analysis does not in itself result in an understanding amongst students as to why particular decisions are made during data analysis. In this particular case, the teaching activity has been modified in order to focus future students' attention upon the reasons why important decisions are made during data analysis. For example, students will be presented with 4 contrasting views as to how a value for the initial rate of reaction should be determined. These are based upon responses that previous students made during the survey. Students will then be asked to discuss these, and decide which was the most appropriate course of action.

By conducting research on the ways in which students attempt teaching and learning activities, and the reasoning that they appear to draw upon in acting, it becomes possible to identify potential difficulties in communication between teachers and students. The teachers responsible for this biochemistry activity were very surprised at the ways in which their students conceptualised data, and relationships between experimentally determined constants and the Michaelis-Menten model. The potential value of research on students' performance in practical work is therefore in identifying new learning objectives for the curriculum in areas previously thought to be unproblematic for learners, and in identifying areas of particular teaching and learning activities about which teachers need to make their own assumptions more explicit to students to avoid miscommunication.

Acknowledgement

The work reported in this paper was partly supported by the European Commission DGXII through the project *Labwork in Science Education* (PL95 2005). The author acknowledges valuable discussions held with members of the LSE project consortium throughout the project. The research was also partly supported by the Economic and Social Research Council (Grant R000233286).

John Leach

References

Bandiera, M., Tarsitani, C., Torracca, E.,Vicentini, M., Dupré. F. and Séré, M-G. (1998) *Teachers' image of science and labwork. Hypotheses, research tools and results in Italy and in France* Working Paper 5, Labwork in Science Education Project. Rome: University of Rome 'La Sapienza'.

Bruner, J. (1962) *Introduction* in L. S. Vygotsky: *Thought and Language* Cambridge, MA: The MIT Press.

Carey, S., Evans, R., Honda, M., Jay, E., Unger, C. (1989) *'An experiment is when you try it and see if it works': a study of Junior High School students' understanding of the construction of scientific knowledge* International Journal of Science Education, 11, (5) 514-529.

Department of Education and Science (1991) *Science in the National Curriculum* London: HMSO.

Désautels, J. and Larochelle, M. (1998) *The epistemology of students: the 'thingified' nature of scientific knowledge* in B. Frazer and K. Tobin (Editors) *International Handbook of Science Education* Dordrecht, NL: Kluwer.

Driver, R., Leach, J., Millar, R. and Scott, P. (1996) *Young people's images of science* Buckingham: Open University Press

Kuhn, D., Amsell, E. and O'Loughlin, M. (1988) *The development of scientific thinking skills* Academic Press.

Koslowski, B. (1996) *Theory and evidence: the development of scientific reasoning* Cambridge, MA: The MIT Press

Leach, J., Millar, R., Ryder, J., Séré, M-G., Hammelev, D., Niedderer, H. and Tselfes, V. (1998) *Survey 2: students' images of science as they relate to labwork learning* Working Paper 4, LSE Project. Leeds: Centre for Studies in Science and Mathematics Education.

Leach, J., Driver, R., Millar, R. and Scott, P. (1993) *Students' characterisations of empirical enquiry* Working Paper 6, Students' Understanding of the Nature of Science Project. Leeds: Centre for Studies in Science and Mathematics Education.

Leach, J. (in press) *Students' understanding of the co-ordination of theory and evidence* International Journal of Science Education.

Leach, J. (1998) *The use of secondary data in teaching abot data analysis in a first year undergraduate biochemistry course* in J. Leach, J. Lewis and J. Ryder: *Learning about the actual practice of science: three case studies of undergraduate labwork from the UK* Leeds: Centre for Studies in Science and Mathematics Education.

Lubben, F. and Millar, R. (1996) *Children's ideas about the reliability of experimental data* International Journal of Science Education, 18, (8), 955-968.

Matthews, M. R. (1994) *Science teaching: the role of history and philosophy of science* London: Routledge

Metz, K. (1991) *Development of explanation: incremental and fundamental change in children's physics knowledge* Journal of Research in Science Teaching, 28, (9), 785-798

Mortimer, E. F. (1993) *Studying conceptual evolution in the classroom as conceptual profile change* Proceedings of the third international seminar on misconceptions in science and mathematics education, Ithaca, NY: Misconceptions Trust.

Rutherford, F. J. and Ahlgren, A. (1989) *Science for all Americans* New York: Oxford University Press

Samarapungavan, A. (1992) *Children's metajudgements in theory choice tasks: scientific rationality in childhood* Cognition, 45, 1-32.

Séré, M-G., Larcher, C. and Journeaux, R. (1993) *Learning the statistical analysis of measurement errors* International Journal of Science Education, 15, (4), 427-438.

Teacher Training Agency (1998) *Initial Teacher Training National Curriculum (Secondary Science): Consultation Document* London: Teacher Training Agency.

Tiberghien, A., Bécu-Robinault, K., von Aufschnaiter, S., Buty, C., Fernandez, M., Fischer, H., Leach, J., Le Maréchal, J-F, Molohides, A., Paulsen, A. C., Pol, D., Psillos, D., Salame, N., Tarsitani, C., Torracca, E., Veillard, L. and Winther, J. (1998) *Science teaching and labwork practice in several European countries* Working Paper 1, LSE Project. Lyon: Equipe COAST, Université Lyon 2.

The interaction between teaching styles and pupil autonomy in practical science investigations
– a case-study

J. R. Watson, J.R.L. Swain and C. McRobbie

The influence of the teaching styles of two teachers on the structure and organisation of investigational lessons is explored. These are shown to influence way in which pupils (aged 12/13) made decisions in different stages of investigational lessons. Each teacher was followed for nine lessons and the actions and activities of the teacher and two target groups of pupils were recorded on video-tape and audio-tapes. The observations in the classroom were supplemented by collecting documentary evidence, interviewing the participants and using questionnaires. The teaching style of the teachers is seen to affect the autonomy of decision-making of pupils.

Introduction

Although practical work and scientific enquiry were well established in the curriculum in England and Wales (e.g. DES/WO, 1995), the introduction of a National Curriculum (DES/WO, 1989) placed an increased emphasis on investigative work. The implementation of investigational work has, however, proved problematic (Donnelly et al. 1993; Donnelly, 1994; Donnelly et al., 1996; Foulds, Gott et al., 1992). The Open-ended Work in Science (OPENS) project (Jones et al. 1992) developed strategies to support learning in investigations practical work and found that it was often difficult for teachers to change their practice. The current study explores one aspect that appeared to affect teaching in investigations: the effect of teachers' existing teaching styles.

Investigations can be seen as problem-solving situations in which students make a series of decisions in order to reach a conclusion. Millar et al., (1994) adapted the APU problem-solving cycle (Schofield et al., 1983) to explore the effects of students' declarative and procedural knowledge on the responses of groups of students to different stages of investigative tasks. They saw the stages of investigations as triggers which stimulate selective recall from students' memories in order to construct responses to the task. The students' responses indicated that students were matching the set task to recalled classroom episodes which Schank and Abelson (1977) called 'scripts'. Scripts were modified by students' procedural (knowing how) and declarative (knowing that) knowledge.

In the current study the approach of Millar et al. (1994) is developed in order to apply it to teaching investigative lessons. The study focuses on the effects of the ways that two teachers structured and organised their lessons, on the decision-making of pupils. In particular it explores the influence of teaching style on how two teachers interpreted an intervention designed to enhance the quality of decision-making in investigations.

Previous studies have identified features in a teaching and learning situation that affect how groups of pupils responded to investigative tasks (Jones et al., 1992; Simon et al., 1992; Watson, 1994). Interactions in groups were affected by the pupils' perception of the task (i.e. which 'scripts' are recalled), pupils' procedural and declarative knowledge and the ability of members of the group to collaborate. Significant aspects of the structure and organisation of the investigation lessons were the structure and timing of parts of the lessons, the organisation of learning activities, the presentation of the investigation, the apparatus supplied, worksheets used by students and the nature of classroom interactions including the arrangement and size of student groups.

Method

Of three teachers initially observed teaching year 8 (age 12-13) classes, two with differing teaching styles were selected for the study. They were observed in nine 50 minute lessons each, in three kinds of lessons: ordinary science lessons; a set of investigative lessons; and, after intervention by the researchers, investigational lessons about the strength of paper chains. The actions and talk of the teacher and two target group of pupils were recorded using two video cameras and seven tape-recorders. The 13 hours of video-tape and 80 hours of audio-tape have been transcribed to produce two paths through each investigations, one for each target group. This data is complemented by field notes, photocopies of all the written work, interviews of the two teachers, interviews of the pupils in the target groups and a pupil questionnaire completed by all members of both classes.

The process of analysis has been an iterative one. Data from the different sources has been analysed and descriptions generated from different sources compared. The description presented is the researchers' interpretation of the transcripts and video-tapes, complemented by the perspectives of the pupils and teachers.

Teaching styles of the teachers before the intervention

Lena and Peter (not their real names) were young teachers with a few years' teaching experience. Both were perceived to be good teachers by colleagues and the researchers. They were enthusiastic, well organised and related well to children. They both saw investigations as giving more autonomy to pupils and Lena also saw them as opportunities to deepen pupils' conceptual understanding.

In whole class work Lena had a friendly and informal questioning style and created the impression that she wanted to hear the pupils' responses and valued them. Pupils thought that 'she had a soft spot for them.' When giving directions she was direct and to the point. In interview she described her role as a teacher:

> ... there is information that they need to know ... the only way you can really teach them is by telling them it... but some of the time I know that ... I can...facilitate, give them some ideas on how they can find things out.

Lessons usually began with a whole class discussion and then most of the time was spent in small groups when Lena focused on exploring pupils' thinking and what they knew already. In investigations pupils said:

> She'd give you, like... a word, and you would have to...build on it...She'd give you like a little clue, and that would make you think further ahead.

The interaction between teaching styles

In spite of the friendly atmosphere created, Lena had firm control over the class.

Peter's lessons tended to be dominated by whole class teaching. He gave clear explanations but tended to spend quite a lot of time explaining exactly what to do in practical activities. When questioning he tried to get responses from around the class, but a few boys dominated. Peter saw his role as a motivator, transmitter of knowledge and to some extent a controller.

> *Peter*: I think the teacher... is sort of the dispenser of information and knowledge... I think to a large extent the teacher needs to be also, if possible, a motivator... and to some extent a controller...

In the words of two pupils:

> S: He likes us... to listen and to concentrate...
> SE: He moulds us!

Peter was prepared to give the pupils some autonomy in group work, but felt that in small groups 'there's no guarantee that they'll be...focusing on the work'. In group work Peter tended to focus on what pupils were doing rather than their thinking. This led to a weak knowledge of pupils' prior experiences. There was a clear distance between Peter and the pupils and they sometimes felt uncertain about asking for help. Although he had good control of the class, it lacked the cohesiveness of Lena's class and occasionally there was some misbehaviour.

Both teachers had used investigations with their classes before the intervention but used the words 'investigation' and 'assessment' interchangeably. Their perceptions of investigations were dominated by the National Curriculum:

> *Int.*: What do you see as your aims in science teaching?... Would you... say that you've got any sort of priorities? I mean, do you think, number one that anything stands out in particular?
> *Lena*: Um, well, I suppose the priority is always to follow the National Curriculum, knowing that they're going to be tested on it at the end of the course.

They viewed investigations as opportunities to assess pupils' procedural knowledge against the criteria described in the National Curriculum.

J. R. Watson, J.R.L. Swain and C. McRobbie

The planned intervention investigation

The intervention took place after the teachers had each been observed for six lessons. The researchers planned an investigation about the factors affecting the strength of paper chains with the two teachers. It would take place over three 50 minute lessons . The main purpose of the intervention was to structure and organise the lesson in such a way that it enhanced decision-making processes in the investigation. It was designed to provide time for specific parts of the investigational process: focusing, planning, obtaining evidence, interpreting and evaluating and further evaluation. The learning focus was, therefore, to be on developing procedural knowledge. Aspects that had were emphasised more strongly in this plan, than in the previous investigation, were focusing and evaluating. Previous investigational work done by the classes had treated investigations as written products for assessment. This has resulted in some pupils using an 'engagement frame' (Millar et al., 1994), seeing investigations as sets of routine procedures. The intervention was planned to shift more pupils towards a 'scientific frame', in which they would be exploring the relation between two variables in a more meaningful way.

The focus of each stage of the investigation was to be supported through interactions with the teacher, a series of worksheets, the materials provided for the practical activities and the pupils' experience of the physical phenomena involved in the practical activities (table 1).

Stage of Lesson	Purpose	Structure	Organisation
Focusing	What is the task about? • Present problem to pupils. • Elicit relevant knowledge through practical work.	Lesson 1 • Short whole class introduction (3 min) • Practical activity (15-20 min) • Whole class discussion (5 min)	Focus on observation: how and why the paper chains broke and making quantitative predictions. Group work: • Pupils work in friendship groups of 2 or 3. • Each group provided with 8 strips of papers, of different shapes, sizes and types. • Teacher discussion with groups

The interaction between teaching styles

Stage of Lesson	Purpose	Structure	Organisation
Planning	What are the key variables and how can they be measured?	Lesson 2 • Demonstration of how to measure dependent variable (force to break chain) (5 min). • Planning in groups (10 min)	• Worksheet 1: Blank variables to help to identify and operationalise key variables. • Worksheet 2: Emphasis on quantitative prediction and explaining predictions. • Apparatus supplied: strips of paper, glue etc. • Teacher discussion with groups: quantitative predictions and fair tests.
Doing	Decide how to make the chains: operationalise independent and control variables. Evaluate values, number of measurements and controls. Decide on repeats. Judge errors	• Whole class introduction (3 min) • Making and testing paper chains (30 min)	• Worksheet 1: Use variables table to record measurements. • Worksheet 2: Quantitative prediction and explaining. • Apparatus supplied: strips of paper, glue etc. • Teacher discussion with groups: problems of operationalisation and evaluation of the quality and number of measurements made.
Interpreting and Evaluating	What patterns can be seen in the data? Are the data reliable and sufficient?	• Small group discussion about patterns (10-15 min) Lesson 3 • Prepare results for whole class review (5 min) • Group presentation of results (10 min)	• Worksheet 3 guides discussion about patterns and quality of data. • Focus of presentations and peer review is on patterns, quality of data and how it could be improved.
Further evaluation	How can the quality of the data be improved? • Measuring techniques • Range of measurements Can quantitative predictions be made from the patterns?	Pupils work in groups to check measurements, or make measurements to evaluate quantitative predictions (20 min)	• Worksheet 4 encourages pupils to extend their investigation. • Teacher works with groups to set targets for extending investigations.

Table 1: *Outline of planned intervention investigation*

J. R. Watson, J.R.L. Swain and C. McRobbie

Investigation Lessons Observed

This section starts with an overview of what happened in the investigation lessons and then the first of the series of investigation lessons is described in detail to illustrate the effects of the teachers' interpretation of the intervention investigation on the pupils.

Overview of paper chains investigation

There were some similarities between the lessons of the teachers. Both concentrated on small group work, reflecting the greater openness of the investigation. In both lessons the pupils were unclear about the learning objectives and focused on surface features such as learning that graph paper is stronger than newspaper, or learning how to make a paper chain:

Int.: What do you think you've learned from doing your investigations?...
R: ... that graph paper is stronger, that green one.
Int.: Right. Is that it?
R: Um...
Int: You spent three lessons doing that, seems a long time to spend finding out that graph paper's stronger.
JA: Yeah, it, and we also found out which, um, which paper's stronger. Not just the graph paper, all of them.

Many pupils approached the investigation as a routine exercise. They saw the worksheets as guiding them through set procedures and many seemed to view satisfactory completion of the investigation as producing a set of completed worksheets. For example, the target group of girls in Lena's class investigated the effect of using different glues to stick the chains. They wrote up their investigation concluding the order of strength of the glues, yet in interview they revealed that they knew that the glue was having no effect on the strength of the chains and that the chain never broke where it was glued. This approach fits well with the teachers' previous use of investigations as 'assessments' in which they used the pupils' written work to assess them.

There were also significant differences. Lena felt it important to give the pupils some responsibility for their own learning whereas Peter revealed a reluctance to hand over control to pupils:

The interaction between teaching styles

...there's always a slight fear of doing that, that you may not get to a particular group early on, and they may have gone a long way down the wrong route...

Lena's class felt that she listened to their ideas and encouraged them to think for themselves whereas Peter's class thought that direct teaching was more important. The atmosphere in the two classrooms was different: both the pupils and Lena thought the pupils worked quite hard in this investigation, whereas Peter's class lacked cohesiveness reflected in some minor misbehaviour by some boys.

Lesson 1 of the investigation

Lesson 1 superficially reflected the planned structure for the investigation for both teachers: a whole class introduction, then pupils working in groups to construct and pull apart paper chains and finally a whole class planning stage.

The focusing stage
Lena began this stage by gathering the class around her and giving out the worksheets for the planning stage. She focused on making observations of paper chains being pulled apart and thinking about why they broke. She quickly modified this emphasis:

Now, what I want you to do them is to PREDICT together, which LINK is going to break first...

The pupils then worked in small groups. The response of the two target groups to the introduction was quite different. The group of boys immediately began constructing their chain, clarifying their understanding of how it was to be constructed. They predicted which link would break first and pulled the chain apart into successively smaller bits and gave a variety of reasons for particular links breaking first. The girls' group started with a lot of off-task talk. Eventually they started sorting and describing their strips of paper. One girl asked the teacher where she should write her ideas. The worksheet was not designed for this, but rather as a table for listing and deciding values of key variables, but they agreed that the girls should write their ideas on a blank part of the worksheet. The girls then began their writing accompanied by a lot of off-task talk. Later the teacher returned to the group and the girls wanted to know whether what they had written was acceptable. The teacher said it was and encouraged the pupils to construct the paper chain, which they then did. This was accompanied by a discussion of whether they were conducting a fair test. They eventually decided that they had to predict where the chain would break, rather than carry out a fair test. This discussion was interrupted by the whole class session on planning and the girls never pulled their paper chain apart.

Peter's introduction focused on identifying key variables. It was not until the end of the introduction that he modified this and told the pupils to focus on how the paper chain would break and why.

After the introduction the pupils worked in groups. The girls group made two predictions of which link would break in the chain. These predictions were based on the position and the thickness of the strip of paper. Although different members of the group disagreed on the predicted effects of both these, no reasons were given to support their assertions. When the girls actually broke their chain, it broke near the middle and the girls' discussion then focused on whether the break was actually at the middle or near it. This discussion was unresolved and drifted into off-task talk. At no time did they discuss why it should have broken where it did. During this time the teacher visited the group briefly on two occasions and on each occasion focused on identifying key variables rather than why or how the chain had broken.

The response of the target group of boys in Peter's lesson was characterised by large amounts of off-task talk and minor misbehaviour. They carried out the practical work required, constructing and breaking the paper chain, but failed to engage in discussion about the problem. The teacher visited the group on several occasions for very short periods of time but they continued their off-task talk as soon as he left and sometimes when he was talking to them. Peter's desire to retain firm control on the transmission of knowledge was apparent throughout the lesson. During group work he made frequent interjections to the whole class trying to focus them on particular aspects of the task, typically beginning with, "Please listen very carefully."

The planning stage
In both classes the planning stage began with a whole class discussion. Lena gathered the pupils around the front and explained how to use the blank variables table in worksheet 1, for planning the investigation. As soon as the class had returned to their seats the girls' group went up to the teacher and

asked for help in filling in the blank variables table, spending about 5 minutes with her. On returning to their seats they were satisfied that they had completed the necessary paperwork and spent the last 5 minutes talking off-task and never pulled their paper chain apart. Meanwhile the boys made a half hearted attempt to fill in the variables table and then spent the last ten minutes chatting among themselves.

Peter introduced the planning stage to the whole class by giving out the planning worksheet and his question and answer session focused on identifying what 'is most important about how the chain might break.' There was no discussion of why these factors affected the strength of the chains or how they might be measured. The pupils then worked in groups writing down the key variables. Both target groups did this in a desultory fashion, spending more time talking off-task than on-task. Paul then had a final whole class session which focused again on what were the key variables.

Discussion

The study shows a strong interaction between the planned intervention and the teachers' prior teaching styles and views of the role of investigative work in the science curriculum. The teachers saw their role as helping students to master the skills and processes listed in the National Curriculum. Millar and Driver (1989) maintain that the skills and processes of science only gain their scientific character through the scientific purpose and concepts in which they are embedded. If the main purpose of an investigation is to produce a written product for assessment, then the skills and processes needed to perform the activity run the risk of becoming routine procedures to be learnt, rather than something with intrinsic meaning. This can lead to pupils working in an engagement frame; seeing investigations simply as activity.

The teachers' view of investigations as 'assessments' meant that they found it difficult to change to an approach which emphasised supporting pupil decision-making and retained a strong emphasis on producing a written account of the investigation. The orientation of the teachers to assessment emerged when they introduced the investigation to the whole class: both teachers gave mixed messages as to what this stage was about, as it did not focus on a procedure that they would normally assess. The focus shifted from how and why the chains would break, to predicting or identifying key variables. This mixed focus was reinforced by the ways that the teachers adapted the organisation of the lessons, e.g. the ways in which worksheet 1 was used and in the teachers' interactions with groups.

The different teaching styles of the teachers affected how they interpreted the intervention. Lena could adapt her teaching style, whereas the approach intended in the intervention required a radical change to Peter's style. Lena saw investigations as opportunities to develop pupils' conceptual understanding, already placed some of the responsibility for learning on pupils and was used to situations where she encouraged pupils to use and develop their own ideas. Peter's lesson, superficially, was closer to the planned lesson in structure but the message conveyed through his interactions with the pupils was that all that was necessary to complete this lesson successfully was to identify the key variables in the investigation. His desire to 'mould' the pupils was apparent in his reluctance to give pupils autonomy and his frequent interjections to the whole class.

The effect of the teachers' structuring and organisation of the decision-making processes of the groups varied. The pupils in Peter's class appeared to take a passive role. Most accepted Peter as controlling the class's activities and carried out the practical activity but only responded to thinking about the task in direct response to the teacher's questions. The decision-making processes remained mainly with the teacher. In Lena's lesson the response was more mixed. The girls group failed to pick up the fact that the investigation was different from their previous 'assessment' investigation and concentrated on completing the worksheet. They had matched the task to the wrong 'script'. In her discussions with the girls, the teacher inadvertently reinforced this view. The boys group, however, which had less contact with the teacher, picked up the intended focus from the teacher's introduction and carried out the focusing stage in a similar way to that anticipated. What is common to all these groups is that they had no clear idea of the learning objectives of the investigation to guide how they responded to the investigation. Pupils were often observed working in an engagement frame, carrying out practical tasks without understanding and without any clear criteria to evaluate their purpose.

Previous work has concentrated on how groups of pupils respond to different tasks in non-teaching situations. The current study illuminates how different aspects of the structure and organisation of investigations may be mediated by the teacher. Effective orchestration of pupils' decision-making in investigative lessons is a difficult teaching challenge. Teaching style was one of the factors that mediated whether the teachers were able to use new pedagogic approaches, but equally important may be the ways in which pupils and teachers view the purposes of investigative work.

References

Department of Education and Science/Welsh Office (1985) Science 5-16: A Statement of Policy. London, HMSO.

Department of Education and Science/Welsh Office (1989) Science in the National Curriculum. London, HMSO.

Donnelly, J.F., Buchan, A.S., Jenkins, E.V. and Welford, A.J. (1993) *Investigations in Science Education Policy: Science 1 in the National Curriculum*, CPSE University of Leeds.

Donnelly, J.F. (1994) Policy and Curricular Change: Modelling Science in the National curriculum for England and Wales *Studies in Science Education* 24 100-128.

Donnelly, J.F., Buchan, A.S., Jenkins, E.V. Laws, P. and Welford, G. (1996) *Investigations by Order: Policy, Curriculum and Science Teachers' Work under the Education Reform Act*. Nafferton, UK: Studies in Education.

Foulds, K., Gott, R. and Feasey, R. (1992) *Investigative Work in Science: a report by the Exploration of Science Team to the National Curriculum Council*, University of Durham.

Jones, A., Simon, S., Black, P.J., Fairbrother, R.W. and Watson, J.R. (1992) *Open work in Science: Development of investigations in schools*, Association for Science Education: Hatfield, Herts.

Millar, R. and Driver, R (1987) Beyond Process *Studies in Science Education*, 14, 33-62.

Millar, R., Lubben, F., Gott, R. and Duggan, S. (1994) Investigating the school science laboratory: conceptual and procedural knowledge and their influence on performance *Research Papers in Education* 9(2) 207-248.

Schank, R.C. and Abelson, R.A. (1977). Scripts, Plans, Goals and Understanding. An Inquiry into Human Knowledge. Hillstate, NJ: Lawrence Erlbaum.

Schofield, B., Black, P., Head, J., Murphy, P. 1983, *Science in schools: age 13, research report no. 2*, London, DES.

Simon, S., Jones, A., Black, P.J., Fairbrother, R.W. and Watson, J.R. (1992) *Open-ended work in science: a review of existing practice*, King's College: London.

Watson, J.R. (1994) Students' engagement in practical problem-solving: a case-study *International Journal of Science Education* 16(1) 27-43.

Is authentic appropriate?
The use of work contexts in science practical activity

Susan Molyneux-Hodgson, Rosamund Sutherland and Anne Butterfield

Students studying a vocational science course in England carry out a substantial amount of practical activity. Their practical work usually involves using contexts drawn from the world of work, for example, the chemical industry, manufacturing and health services. The rhetoric is that students will be more likely to learn and enjoy science when they take part in the same types of activity as scientists in the workplace. We need to understand whether this approach is the most appropriate way for students to learn and thus a research project has looked at the premises on which this learning approach is based and what happens in classroom practice.

Introduction and background

Introduced in 1992, as the result of a Government policy paper concerned with the skills base and competitiveness of the UK workforce, the General National Vocational Qualification (GNVQ) was intended to bridge a gap between wholly work-based qualifications and school or college-based qualifications. The GNVQ was aimed at providing a broad educational base in a vocationally-oriented area and would give students a route to either employment or higher education. GNVQs are studied within educational institutions (e.g. schools and Further Education colleges) although the emphasis within the course is on the workplace and vocational area of interest.

The GNVQ in Science was introduced fully in 1994 and is studied predominantly in colleges of Further Education by full time students, usually aged 16-19 years. Drawing on the vocational intention of the course, students studying science develop their skills, knowledge and understanding of science through experiencing the types of activity that scientists carry out e.g. testing to gather data; analysing things; solving problems (…)   (NCVQ, 1996)

The emphasis then, in the rhetoric, is on students carrying out scientific activity. As far as possible, this activity is expected to be conducted within appropriate vocational contexts, so for example, science topics related to cell and tissue biology could be learned through the context of studying a cervical cancer screening programme (Nuffield, 1994).

As well as activity, a further key tenet of the GNVQ is that of 'student as an independent learner'

An important aspect of the GNVQ curriculum … is that students take greater responsibility for their own learning. This … allows the use of flexible and efficient learning modes, and makes effective use of teacher time and physical resources.   (NCVQ, 1995)

The combination of intentions of learning through science-like activity and of independent learning means that in practice, students studying GNVQ Science engage in a tremendous amount of practical work. That this practical activity is situated within work-based contexts is of prime importance to the qualification's framework. Note that the aim for this practical activity is not that students learn any 'scientific method', but that they learn science by doing what people who use science, do. Note also that this 'doing' is not the activity of research scientists but of people such as nurses, health and safety officers, engineers, laboratory technicians and environmental health officers.

Theoretical background

It is probably the case that practical-based learning of science, construed within a limited view of laboratory-based practical work, is not the most effective way to learn science. It may have been within this context that Osborne stated, "the doing and learning of science are not the same thing" (1997, p.61). Brown et.al. (1989) presented a strong case to demonstrate that the separation between the categories of knowing and doing is fundamentally flawed, and they argued for the centrality of activity to learning

> the activity in which knowledge is developed ... is an integral part of what is learned ... any method that tries to teach abstract concepts independently of authentic situations, overlooks the way understanding is developed through continued, situated use. (Brown, 1989)

From this perspective the GNVQ approach to learning, aiming to promote learning through the use of authentic situations, seems promising.

There is a move towards increasing students' awareness of the work environment in the UK with the QCA[1] advocating an increased role for the use of work contexts in all schooling (QCA, 1998), equating this with more "purposeful learning". The view seems to be that learning through such contexts will produce knowledge which is more useful in practice and will be more 'transferable'. Indeed, a move towards the use of 'authentic' activity in school and college education appears to be gaining popularity in general (for example, Berenfield, 1997; Hall, 1997; Roth, 1995; Woolnough, 1998).

One strand of thinking emanating from the field of sociology of science involves a view of the learning of science as an enculturation into communities of practice (for example, Pickering, 1992). Ideas around communities of practice have also been explored by Lave & Wenger (1991) who offered the construct of 'legitimate peripheral participation' as an explanatory mechanism for understanding people's increasing involvement in a community of practice and movement from outsider to full participant. These perspectives seem potentially fruitful given the emphasis within GNVQ on participation in scientific activity and development of the independent learner.

Research setting

We conducted a research study to investigate practical activity within the GNVQ framework, as part of a larger project between October 1995 and December 1997 (Molyneux-Hodgson et.al., 1998). In this paper, we will focus on the questions, what are the roles of work contexts in students' practical activity?; how does this practical activity influence students' learning?; and what can be learnt from this work-context approach which is of relevance to science learning in general?

---

[1] The Qualifications and Curriculum Authority, responsible for curriculum, assessment and qualifications at all levels of education and training in England.

Is authentic appropriate?

The students we worked with were studying at the Advanced level, that is, at a level equivalent to the normal University entrance qualification in the UK, the 'A' level. The GNVQ at Advanced level consists of 8 compulsory units plus 4 optional units and 3 'core skills'[2] units and this would take up the majority of students' time over a two-year period. Students gain the GNVQ qualification by providing evidence that they have fulfilled lists of 'performance criteria'[3] which they usually do through production of written assignments. The teacher's role is not usually 'to teach' but to provide opportunities for students to collect evidence of fulfilling the criteria. We collaborated with two Colleges of Further Education in different cities in England, involving 27 students (aged 17-19 years) and 7 members of teaching staff. The students were studying full-time and had chosen to study GNVQ rather than 'A' levels.

The data collected included field notes on observations of teaching sessions, discussions with teachers, a series of semi-structured individual student interviews and collection of student assignments and other written work. Using the diversity of data collected, case studies of students were developed, alongside case studies of particular science topics. Here the word 'topics' is used broadly and can mean, for example, a micro-focus on converting between units of measurement, or an overview of chemistry. These case studies take into account the ways in which a topic is presented to students; actual student practices and issues such as comparing student practices with canonical scientific interpretations. Analysis of the data involved working with the interaction between student action, written texts and verbal statements and was informed by the theoretical strands discussed above.

Student voices

At this point we will give the science students the a voice, in order to gain their verbal opinion on the GNVQ approach. They are generally positive about the widespread use of practical work, for example:

---

[2] All GNVQ students must study the core skills of communication, information technology and application of number, alongside their main studies.
[3] The performance criteria list what the student should be able *to do*, in contrast with a traditional subject syllabus which states content to be covered.

> 12.70. Carol: ... we get to do more experiments, more practical things which is easier and you can get a better understanding of what you're doing rather than just learning it (by rote), ... and you get to see why.

They were also keen on the idea of independent learning:

> 12.24 Di: I think it sinks in a bit more if you actually have to go and find it out for yourself. You research things for yourself rather than just sitting there taking notes.

But they held mixed views about the use of work-contexts:

> 12.250 Di: Pointless, really.
> 12.96 Carol: it's difficult ... design an experiment to show whether it would actually work ... when it's scaling things up like that. I mean it's useful to know how to scale things up ... so I think it does have its uses.
> 12.186 Nat: Somewhere along the line you have to fit that in. But I tend to look at the range (the performance criteria) ... rather than the (context).

However, as discussed above, what students say is only a component part of a complex picture and often their conceptions of ideas such as independent learning are somewhat different from our own, thus their voices need to be interpreted within a wider framework.

The roles of work contexts

What are the work contexts that feature so prominently in the GNVQ rhetoric and which Di (see above) was unimpressed by? It is accepted by tutors delivering the GNVQ courses that they should present students with opportunities for learning which are situated within work contexts. Both colleges we worked with used a mixture of resources to present ideas to students. As an example of this, a set of tasks designed to complete one of the science modules (Unit 3, Obtain New Substances) is presented below:

- You are providing an advisory service to the chemical industry .. plan and carry out the preparation and purification of three substances.
- A sudden increase in the demand for aspirin requires a review of the preparative route currently in use and the consideration of possible alternative routes.
- Nitration is a commonly used step in multi-step preparations of organic compounds ... carry out the nitration of the methyl ester of benzoic acid.
- There are many instances of organic compounds extracted from plant material ... obtain a sample of limonene from orange peel.

(BTEC 1994)[4]

This resource pack stated that three of these tasks were "placed in a commercial context" (ibid.) although which three tasks is not immediately apparent. It is unlikely that these short statements are sufficiently rich to either engage students in purposeful learning or provide tutors with enough information to support student learning, especially if these are work situations that the tutor has not experienced. Although more information is provided in the resources than that given above, it is questionable whether the examples provide work contexts which would enable students to engage in practice which could in any way be construed as authentic.

One consequence has been that these suggested work contexts are not used to situate practical activity. Indeed our observational study found that very often the chemistry tutors did not present students with *any* work contexts to structure learning. We did not investigate why this was the case but possible interpretations might be that they could not find appropriate contexts to use, or they considered chemistry work contexts did not make sense in educational settings, or for some other reasons. What usually happened in practice was that the chemistry tutors asked students to conduct some experiments which allowed students to demonstrate particular skills or collect data on which certain manipulations could be carried out. For example, students carried out an esterification reaction, without reference to the world of work, in order to learn about chemical equilibrium and factors effecting reactions, whereas available resources used the piloting of a batch reactor process (Nuffield, 1994), photography and anaesthetic manufacture

---

4   BTEC was one of 3 bodies which awards GNVQs. This awarding body's resources were the most-widely used in both Colleges we worked with.

(Nuffield, 1995) to situate these topics. Thus students engaged in fairly traditional school-chemistry-like practical work the majority of the time and work contexts were rarely mentioned.

The four work contexts listed above are reminiscent in purpose of the 'everyday' contexts used in mathematics education which are meant to make mathematics more accessible to learners through situating problems in non-mathematical contexts, for example, the use of a grocery shop situation to introduce ideas of algebraic equivalence (SMP 11-16, 1984). This contextual approach is being criticised within the mathematics education world for the effects this approach is having, for example, on students' algebraic understanding (Royal Society and JMC, 1997). In a similar way, the chemistry work contexts could be an attempt to make chemistry more accessible, through illustrating applications. However, this is not the main intention of their inclusion with the GNVQ framework; the contexts are intended to be more than just examples of science applications.

Work contexts in action

Unlike the chemistry examples discussed above, work contexts were used overtly to situate much of the learning of physics-related ideas. As an example, we will consider the case of Unit 2 – Investigate Materials and their Use. The work context suggested by the BTEC resource pack for this unit was a manufacturing industry context. The notes for the Unit assessment read:

> In recent years the electric kettle has been redesigned .. there is increased competition between manufacturers ... your company is considering whether to market an electric kettle. It will be necessary to use a suitable material for the heating element ... market research has shown a level indicator (must be incorporated) ...

In addition to this information, students were given a handout by their tutor stating that the overall goal was, "to produce an electric kettle of capacity 2.5 litres".

The tasks students were asked to carry out were written on the handout and included:

Is authentic appropriate?

- you are a member of a design team – produce a 'product design specification' and send it to the Materials Section of the company
- as the materials specialist, determine material properties and send a report to the design team
- imagine you have the design team in the lab, carry out a modification to a material

The 'distance' between the work activity that might be carried out in industry and the suggested activities for students given above, is not particularly large. People working in industry would indeed carry out experiments on material properties for example and these procedures would probably resemble the practical activity students were being asked to carry out. So here, it might be conjectured that students would keep the work context in mind as they worked and the context might structure their activity in some way.

Analysis of the detailed criteria for this unit showed that the students were being expected to learn a great deal of science, including,

learning about materials (metals, polymers, ceramics);
their properties (Young's Modulus, resistivity, specific heat capacity);
how these related to material structure and bonding;
their usage (in construction, domestic use);
their availability, cost, fabrication, environmental impact;
and how they can be modified (alloying, heat treatment, drawing).

The previous paragraphs describe an intention, but what did students actually do? In practice, the students performed an experiment to determine Young's Modulus of a length of copper wire, another experiment to find resistivity of wire, another to find the refractive index of glass and one to determine the specific heat capacity of aluminium. No explicit mention was made during the experimental sessions about the work context and no student referred to the context in their written assignments. So what role was the work context playing here? It is the case that the separate experiments on different materials can be related back to 'the kettle' context, for example, the refractive index of glass could be important in relation to the 'level indicator' for the kettle. However, our analysis of students' work points to the context being used as an excuse to conduct and learn traditional experiments and content, rather than an intention to engage students in any meaningful way with legitimate scientific activity. The prevailing view seems to be that there are significant and exemplary events which students must replicate, for example, doing a Young's Modulus experiment, and a context is found afterwards to give this replication some credibility. This is in contrast to an approach which starts from studying scientific activity in the workplace and extracting salient situations for students to work through.

Note that in many ways the students take the work contexts in their stride,

> I2. 256 Di: ... you're someone at a company designing a kettle and you've got a title and that. And then later in the assignment you're someone else over another department ... it's silly.

Independent learning in practice

The following extracts are taken from field notes of an observation of the session in which students determined the Young's Modulus of copper wire,

The teacher had designated specific times for the students to carry out research into what they would need to do to find Young's modulus. This planning aspect was graded towards their GNVQ. The amount of planning students carried out varied greatly, with one student having produced a page full of background notes, another having designed a results table and the majority of students arriving at the session with nothing prepared on paper.

The above notes exemplify 'independent learning' in practice. It is important to emphasis, that the students were being given control to take decisions about what was important and what was not important with respect to the science work. Many of the students took this responsibility seriously but it is unlikely that they would have sufficient resources to enable them to make appropriate decisions.

The tutor *knew* what he wanted the students to do during the practical sessions, it is just that the GNVQ system is such that students have to discover this for themselves. When students arrived for the sessions, the tutor provided equipment for the practical work which effectively constrained the students into one way of determining material properties. In effect, the tutor 'got his own way' and the status of the students preparatory work became unclear. A specific example of the tutor knowing what was important and students not, was the idea of the validity of Hooke's Law. The teacher had

Is authentic appropriate?

"expected" the students to come across the canonical stress against strain graph in their planning and was surprised when many students set about calculating Young's modulus without plotting anything on a piece of graph paper. One of a pair of girls working at the front of the class, who had done a lot of study prior to the lesson, set about determining Young's Modulus using a completely incorrect method. We have to question how these students are intended to gain science understanding under these circumstances.

The interactions of work contexts and …

There are many other issues at stake besides the ongoing practical work. For example, the GNVQ assessment regime means that students spend a large amount of time managing and administering their learning through portfolio construction[5]. Constraints on resources (including money, space) mean that practical work usually takes place within traditional laboratory classrooms using equipment purchased to allow a more traditional science learning agenda to be followed. This is to say, the situation within which the students are working is very complex and it is perhaps unsurprising that work contexts did not figure strongly in student activity and that often the practical work ends up as a set of conventional practicals which could be observed in many 'A' level classrooms. The most critical point to note is that the quality of science learning is questionable. It is unclear what the students *were* learning and their written work and interviews suggest that it was not the intended science that was being learned.

There were occasions when students did make reference to the work contexts that they had been expected to learn through. However this was usually in the interview context and related to our probing of ideas rather than the students spontaneously making reference in their written assignments. One example of this were comments made by a student being interviewed about his assignment on 'Bottled Water'. Luke referred to the context of a bottle factory throughout his interview, but appeared to have little understanding of the science that this context was meant to have supported him in learning. The bottle factory was another example where the 'distance' between the work context and actual experimental work was not large. The

---

5 A new assessment regime is currently being piloted in recognition of the overburden on participants.

bottle factory, with a conveyor belt and bottles to be filled with water, was intended to support learning of ideas such as velocity, acceleration and fluid flow. The context was relatively easy for to students to visualise as Luke illustrated when he stated what he expected conditions in the factory to be like,

> I2. 211. Luke: Yeah. But we should have just done that all in one 'cause .. the conveyor belt doesn't start off at the slow speed, speeds up and stops.
> I2. 212. Int: How do you know?
> I2. 213. Luke: Well it'd sort of all run at the same speed I would have thought.

Although Luke makes reference to the work context with respect to his own work, his everyday understanding of what a bottle factory might be like was at odds with his practical experiences in the classroom. This presents a further factor for consideration of the appropriateness of work contexts to situate science learning. First-hand experience of the workplace, or personal notions of such places, will be brought to bear by the students and may act in support of, or against, the contexts being used in the classroom.

Summary

Students enjoyed the course but their practical activity was 'detached' from the work contexts intended to give structure and support to their learning. In some ways the role of the work contexts could be considered as 'window dressing'. Students were aware of the limitations of the work contexts and did not always appreciate their contrived nature. Even where the distance between work context and expected student activity was not large, the work contexts gave way to apparently traditional practical activity.

The students were motivated to learn (work contexts as motivator?) and enjoyed finding out for themselves rather than being told. However one consequence of this independent learning was that tutors lost control over what the students had access to. Students are unlikely to develop appropriate scientific understandings when guidance on what is important to pay attention to, is lacking. Students will learn in all contexts and situations, the question remains what is it that they are learning here? In the opening paragraphs of an introduction to the GNVQ Science course, emphasis is given to students' learning through "the types of activity that scientists carry out". The fact that students did a lot of scientist-type activity does not guarantee that the students will learn the prescribed science content.

Although the GNVQ approach of incorporating workplace contexts to support learning seemed promising, the practices we observed were less so. The potential of the GNVQ approach to integrate the learning of skills, content and scientific ways of knowing, is not being played out in practice. The reality appears to be that many factors are present, interacting, so that the institutionalised product does not reflect the pleasing rhetoric. That is, this vocational science does not provide for students' enculturation into a community of scientific practice.

A question, yet to be answered, is to what extent work contexts can ever be meaningfully incorporated into education institutional settings for the learning benefit of students?

Acknowledgement

We want to acknowledge The Leverhulme Trust for funding the project 'Mathematical Practices of GNVQ Science Students', Oct. 1995-April 1998.

References

Berenfield B (1997) 'What is Authentic Science for students? The Global Lab Experience', presented at American Educational Research Association annual meeting, Chicago, March 1997.

BTEC (1994) *GNVQ Assessments: Science, Advanced Mandatory Units.* London, UK: BTEC.

Hall R. & Torralba T. (1997) 'Bringing videographic images of Design-oriented Work into Middle School Classrooms', presented at American Educational Research Association annual meeting, Chicago, March 1997.

Lave J. & Wenger E. (1991) *Situated learning: Legitimate peripheral participation.* Cambridge UK: Cambridge University Press.

Molyneux-Hodgson S., Butterfield A. & Sutherland R. (1998) 'Mathematical Competencies of GNVQ Science Students: The role of computers', Final Report to The Leverhulme Trust, available from Graduate School of Education, University of Bristol.

NCVQ (1995) *GNVQ Briefing: Information on the form, development and implementation of GNVQs*. London, UK: NCVQ.

NCVQ (1996) *GNVQ: Mandatory units and test specifications for Advanced Science*. London, UK: NCVQ.

Nuffield (1994) *GNVQ Science Assignments (Advanced)*, Nuffield Science in Practice, London, UK: Heinemann.

Nuffield (1995) *GNVQ Science (Advanced)* textbook, Nuffield Science in Practice, London, UK: Heinemann.

Osborne J. (1997) Practical Alternatives, *School Science Review*, 78, 285, pp 61-66.

Pickering A. (1992) (ed.) *Science as Practice and Culture*, Chicago, IL: University of Chicago Press.

QCA (1998) 'Work-related learning at key stage 4', *ON-Q*, Issue No. 2, London: QCA.

Roth W-M. (1995) *Authentic School Science: Knowing and learning in open-inquiry science laboratories*, Dortrecht, Netherlands: Kluwer.

Royal Society (1997) *Teaching and Learning Algebra pre-19*, London, UK: The Royal Society and Joint Mathematical Council.

Seely Brown J., Collins A., & Duguid P. (1989) Situated cognition and the culture of learning, *Educational Researcher*, 18, pp. 32-41.

SMP 11-16 (1984) *Algebra*, Cambridge, UK: Cambridge University Press.

Woolnough B. (1998) Authentic science in schools, to develop personal knowledge in J. J. Wellington (Editor) *Practical work in school science: which way now?* London: Routledge.

Section 3

Practical work and teaching scientific concepts

Introduction

One of the central activities of school science is to introduce ideas to students that have been developed in the scientific community to explain the natural world. The four papers in this section each address how practical work can be used as a vehicle for developing students' understanding of scientific concepts.

The first two papers in this section are written by researchers from the same research group. Tiberghien presents an epistemological account of the nature of modelling in physics, and the implications of this for physics education. She then explains the rationale behind a teaching sequence addressing the power concept in terms of this view of modelling, identifying clear learning aims for students in terms of the physics knowledge to be taught, and the students' preinstructional knowledge. Factors influencing the effectiveness of particular teaching activities are then identified, and the learning aims of the teaching are developed and refined. Le Maréchal draws upon a similar account of the nature of modelling to discuss issues involved in chemistry education. He goes on to illustrate and explain some of the difficulties that students encounter when introduced to basic chemical concepts. These kind of careful studies of the teaching and learning of specific subject matter yield valuable insights for work on pedagogy and curriculum development.

Gagliardi, Grimellini Tomasini and Pecori describe a teaching sequence involving practical work, with the aim of introducing some basic Newtonian concepts to primary school students in Italy. The use of practical work is rare in primary classrooms in Italy. An interesting rationale for the teaching is presented.

In the final paper of this section, Eskilsson presents preliminary findings from a longitudinal study of the influence of practical work on Swedish students' understanding of the transformations of matter. In common with Tiberghien and Le Maréchal, Eskilsson draws a distinction between the world of phenomena and the models used to explain phenomena in science. However, the teaching interventions used in Eskilsson's study are not designed specifically to address learning aims identified by the researcher.

John Leach

Labwork activity and learning physics – an approach based on modelling

Andrée Tiberghien

Labwork activity can have different objectives. This study is focused on the role of labwork situations in learning the relations between the material world of experiments, and the conceptual world of physics. The breaking down of knowledge into these two worlds is discussed. Two labwork sessions included in an energy teaching sequence at the upper secondary level are analysed. The first session is at the beginning of the teaching sequence; it involves a modelling activity based on a qualitative introductory theory. The second session is more typical labwork in that students have to make measurements and to find a relation between physical quantities. The results of the first session show how students first establish simple relations between a concept and an object or an event, and then are able to create more complex relations where the experiment is broken down in an unusual way. The students are able to see the experiment "through the eyes" of the physics theory, and not only with their everyday knowledge. In the case of the second session involving quantitative aspects, more sub-levels of the two main worlds are necessary to analyse students' activities. The results show that students mainly construct two types of relations. The first one corresponds to those internal to the theoretical world. This type of relation occurs between the concepts made explicit within the students' natural language, within mathematical formal language, and within the numerical relations coming from measurements. The second type corresponds to relations between concepts and elements of the experiment which are established in a qualitative form. The third type of relation, involving elements of the experiment and quantitative aspects of the model, was seen more rarely. These results suggest that labwork sessions should be carefully designed in order specifically to foster qualitative and quantitative aspects at the level of the physics concepts, the links between qualitative and quantitative aspects and their relations with experiments. Each task should not demand too many different cognitive activities at the same time because students only use a few at once.

Introduction

The fact that labwork activity often has global objectives seems to be internationally agreed upon. The global objective of labwork activity, "to relate theory and practice", is largely recognised by most of the science teachers in several European countries (Welzel et al. 1998). Or, as Lunetta (1998) claims, laboratory experiences have been used for over a century to promote central science education goals including the enhancement of students' conceptual understanding, their scientific practical skills and problem solving abilities, and their interest and motivation.

These approaches use words like 'theory' and 'practice', or terms such as 'scientific concepts' and 'practical skills', as a way of specifying knowledge. Such specifications implicitly or explicitly make categorisations of knowledge. We think that a careful choice of such categorisations should be made in relation to learning hypotheses. In particular, most studies show that experts have a very integrated knowledge of physics, whereas the learner has separate pieces of knowledge. For example, Reif and Allen (1992) show that experts' knowledge involves underlying thought processes in using scientific concepts. At the level of the expert, conceptual knowledge is strongly interrelated with very diverse abilities and skills relative to the elements of experimental situations. However at the level of learners, knowledge is much more fragmentary in the sense that such relations between different pieces of knowledge are not established. In this paper, it is assumed that an important aspect of physics learning is to establish meaningful links between pieces of knowledge. If we intend to study the way the learner constructs this integration, it is necessary to categorise knowledge in order to be able to identify the links established between the categories. The choices made, based on modelling activity, are presented below.

Modelling activity

The choice of modelling activity to analyse teaching situations and students' activities was elaborated from two perspectives: disciplinary physics and learning theory.

Andrée Tiberghien

Physics perspective

The example of Bunge's analysis on the construction of the model of 'bands' in solid state physics illustrates our view of modelling in physics (1973).

> The current theory of the solid state was founded by Bloch four decades ago. Bloch's master idea was to apply wave mechanics (a generic theory) to a simple model of the crystal. The constituents of this model object are a set of fixed points representing an atom each, and a bunch of electrons (or rather model electrons) wandering among the fixed centres. [...] This model object is next conjoined with the vast framework of quantum mechanics. In course of the computations some additional mathematical simplifications may have to be introduced. However, the results are frequently in agreement with the experimental data, which suggest that a nearly true image of real crystal (a non pictorial image to be sure) has been built. Thus, although initially one does not postulate any difference among conductors, semi-conductors, and insulators, this partition is obtained upon analysing the distribution of energy levels (or rather bands). (p.97-98)

```
        ┌─────────────────────────────────┐
       (  Theoretical framework            )
       (  Model within quantum mechanics   )
       (  Model of crystal                 )
        └─────────────────────────────────┘
                      ↕ In agreement with
        ┌─────────────────────────────────┐
       (  Experimental data                )
       (  Crystal                          )
        └─────────────────────────────────┘
```

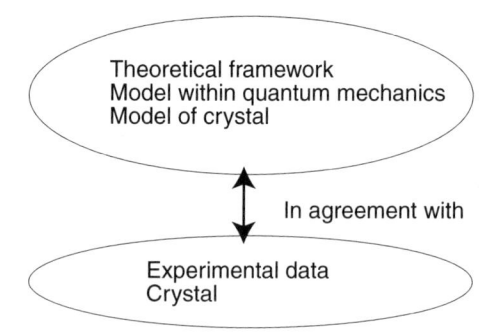

Figure 1: *Categorisation based on Bunge's analysis of Bloch's solid state theory*

Bunge distinguishes the constructed solid state theory which includes the theoretical framework, the quantum mechanical model, the model of crystalline structure, and experimental data. The prediction emerging from the theory and models fits with the experimental data (figure 1).

Learning perspective

When a person or a group of people make an interpretation of, or a prediction about, the material world, a modelling activity is usually involved. This modelling activity involves the two worlds of theory/model on one hand and object/events on the other.

However, research work on students' conceptions suggests that one of the principal difficulties in learning physics is relating these two worlds (figure 2). These two main categories, theory/model and objects/events, allow us to study physics learning by focusing on the difficult aspects of relating knowledge in these two worlds.

This categorisation is different from that usually made, in that abilities and skills such as using algorithms or heuristics are necessary to deal both with theories and with experiments. Similarly, both worlds can include declarative and procedural knowledge (figure 2).

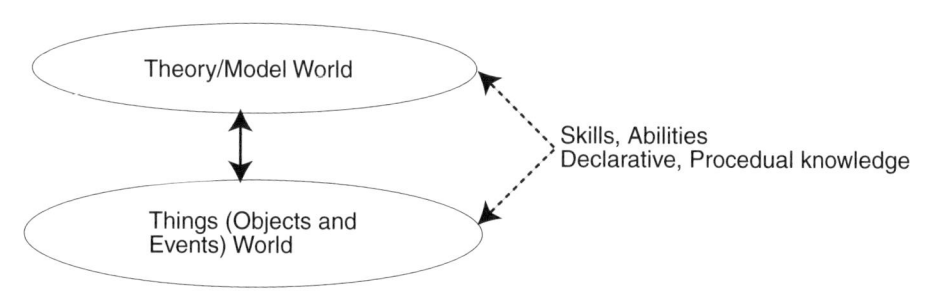

Figure 2: *Categorisation of knowledge based on a modelling activity*

Let us note that the knowledge to be analysed is involved in oral, written or gestural communication and that the analysis is based on lexical and semantic criteria.

In the following sections, a teaching sequence and the associated activities for students is presented, with an analysis of knowledge based on the modelling categorisations. This teaching sequence involves two labwork sessions, and aims at developing relations between these two worlds. The first labwork session involves qualitative modelling, and the other, quantitative.

Andrée Tiberghien

Design of the teaching sequence

This teaching sequence on energy takes place at the upper secondary school level in France, in the stream for students following the scientific route (age 16-17). It is the first time that energy is introduced in the physics curriculum. The research focused on the first three teaching sessions. In their design, the aspects of physics knowledge presented to the students, and those which therefore have to be conceived or constructed by the students, are specified (Tiberghien, 1996). In the case where physics aspects are introduced, the aim of the teaching sequence is that students construct a meaning for these aspects.

- Session 1 aims at creating an intellectual need for a new energy model. The students have already been taught about electricity; they know that when an electrical circuit, like a battery connected to a bulb, is closed, the current flows. This knowledge does not, however, explain the fact that a battery can become 'flat'. A perspective in terms of energy is necessary to explain this phenomenon. This session will not be discussed in detail in this paper.

The initial 'seed' of a theory/model of energy is given to the students between session 1 and session 2 (table 1).

- Session 2 aims at constructing a meaning for the 'seed' of the theory/model of energy. The hypothesis is that to construct a meaning for this energy model, students have to establish relations between this theoretical construction (the 'seed') and the objects and events of the experiment.
- Session 3 aims at refining the theory/model by introducing quantitative aspects and at enlarging the field of applicability. The energy model is developed on the initiative of the students who have to introduce the notion of power.

Theory (seed)	Model (seed)

Energy can be characterised by :
* its *properties*:
– *Storage*
The reservoir stores energy
– *Transformation*
The transformer transforms energy
– *Transfer*
Between a reservoir and a transformer, or between two reservoirs, or between two transformers, there is transfer of energy. The different modes of transfer of energy from a system to another one are :
– <u>by work</u>,
There is transfer of energy under the form of mechanical work when there is movement of an object or of a part of an object during an interaction, under the form of electrical work when there is an electrical current (displacement of charges)
– <u>by heat</u>,
– <u>by radiation</u>.
Energy can also be characterised by :
* a fundamental principle of conservation
Energy is conserved whatever the transformations, transfer and forms of storage
	To build an energy chain
	* the drawn symbols are to be used:
	
		res.	for reservoir
			
			   ———▶ for transfer
			
			   (tr.)
			
			by indicating :
			– in each rectangle the system corresponding to the experiment;
			– under each arrow the mode of transfer;
			
			by putting
			– an arrow by the mode of transfer.
			* the following rules are to be used:
			– a complete energy chain starts and ends with a reservoir;
			– the initial reservoir is different from the final reservoir.

Table 1: *The "seed" of the theory/model on energy given to the students.*

Andrée Tiberghien

Labwork session with qualitative modelling

A short summary of this part of the research is presented in this section of the paper (for further details, see Tiberghien and Megalakaki, 1995). In this session, the students work in pairs on three tasks, each of which has the same problem statement: "By using the 'energy model' (given previously to the students, see table 1), build the energy chain corresponding to the experiment. In the first task, the experimental equipment consists of a bulb, two wires, and a battery. In the second task, the equipment consists of an object hanging on a string which is completely rolled round the axle of a motor (working as a generator). A bulb is connected to the terminals of the motor. When the object falls, the bulb shines (figure 3). In the third task, the equipment consists of a battery connected to an electrical motor. An object is hanging from a string, attached to the axle of the motor, which is completely unrolled before the motor is started. A correct solution was presented to the students after the first task.

In this session, with the help of specific vocabulary (table 1), a symbolic representation, and the energy chain (see an example figure 3), the students are introduced to some basic ideas about the concept of energy. This introduction is compatible with simple causality: the first reservoir can be associated with a cause, and a *single* variable (i.e. the energy) transfers the cause towards a transformer or a reservoir where the effect takes place. The chosen experiments introduce an experimental field where transformations of energy are involved; it is not limited to mechanical phenomena.

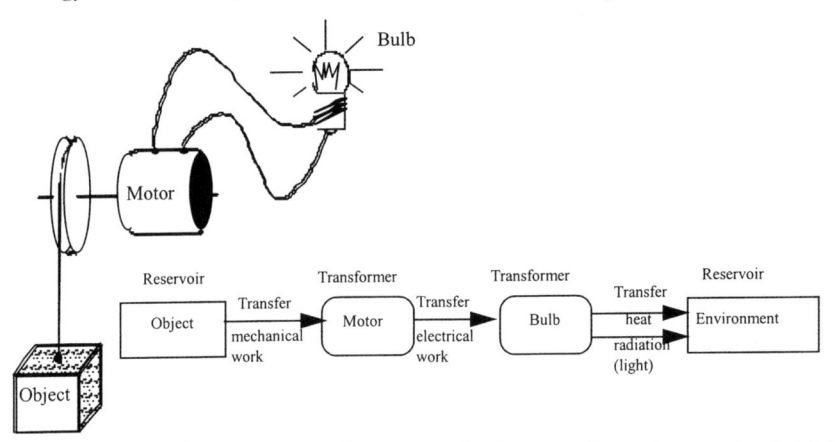

Figure 3: *Example of an energy chain given for the second experiment of task 2 'objects falling'*

Framework of the analysis and discussion of results

The analysis of the whole transcription of six students' dialogues for the three tasks was based upon the meaning of the students' propositions. Several relations were defined in this analysis, and two of them are presented in this paper.

Direct Relation
There is a direct relation when one element of the object/event world is put in correspondence with one element of the theory/model world (figure 4). Example: F-L (during the task 'battery-bulb', dialogue turn n° 22): *'the reservoir stores the energy / thus it is the battery / in the battery there is energy / OK?'*

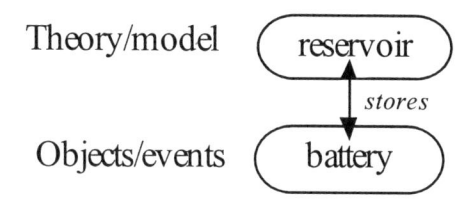

```
Theory/model    ( reservoir )
                     ↕ stores
Objects/events  ( battery )
```

Figure 4: *Direct relation between the two worlds*

Complex relation
There is a complex relation when several elements, or a relation between elements, from one world are put in correspondence with one element, or several elements, or a relation between elements, from the other world. Example: P-F (during the task 'object falling', dialogue turn n° 124 to 126):

P: *I would have thought that the reservoir / that would be the motor plus object together and*
F: *why the motor plus object (?)*
P: *the motor plus object makes the motor run, and after we would put the bulb/ and after we would put the environment.*

Andrée Tiberghien

Theory/model — reservoir

Relation between
- an element of the model
- a relation between two elements of the object/event world

Objects/events — motor object
object makes the motor run

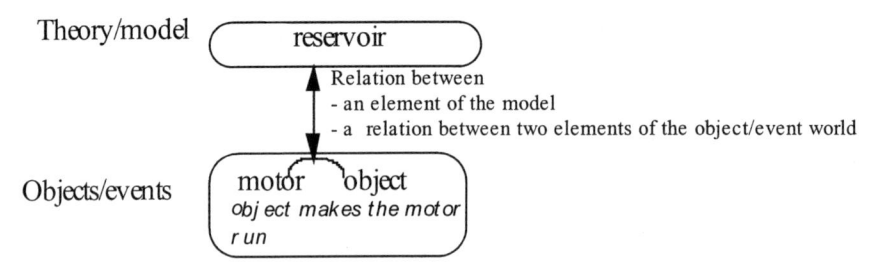

Figure 5: *Complex relation between the two worlds*

The results show that the number of direct relations was more prevalent on the first task, whereas the number of complex relations was more prevalent on tasks 2 and 3 (figure 6).

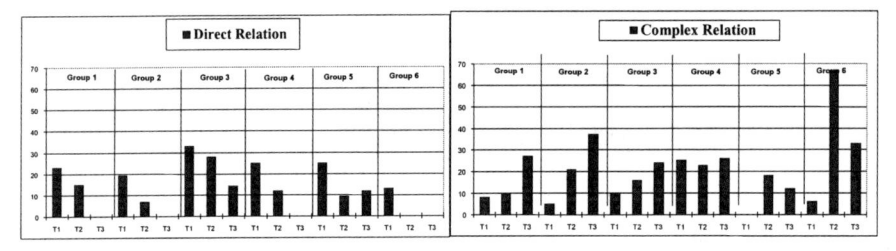

Figure 6: *Percentage of number of direct and complex relations established by six students' groups for the three successive tasks (T1: battery – bulb, T2: object falling, T3: object rising)*

A detailed analysis of the protocols shows that complex relations can happen when the students have acquired some knowledge of the relevant theory. These relations allow students to reinterpret the experiment; this reinterpretation is needed to make the correspondence between the physics concepts and the objects or events of the experimental setting easier. Consequently, when the students do not spontaneously read the experiment in terms which directly fit with the taught physics theory, they have to process a lot of relations between concepts and elements of the experiments before being able to make meaningful simple relations. In the first task, few students reinterpret the experiment as they should have done (2 groups out of the six groups). As a matter of fact, the very familiar battery–bulb experiment must be "viewed" differently in light of the energy model. A new element, the environment, has to be taken into account (for the physicist it is the universe, in order to have a system where energy is conserved). Only in the second and third tasks, did the students begin to be able to use theory to reinterpret the experiment. All six groups used theory correctly, at least in part. Four groups constructed a correct chain during the second task and five groups constructed a correct chain during the third task. The students were able to go back and forth between theory and the experiment. For example, the rule of having a reservoir at the end of the energy chain leads them to ask themselves for the last task 'object rising' a question such as *'well, if there is a reservoir, the energy has to go somewhere'*. Students establish complex relations when they are able to see the experiment with the 'eyes' of the theory. In our case, they can interpret the experiment in terms of reservoir, transformer and transfer of energy.

Labwork session with quantitative modelling

The design of the session is presented first, followed by the results obtained and their analysis. All data were collected and their categorisation elaborated by K. Bécu-Robinault (1997a, b) for her doctoral dissertation; a specific analysis is presented here.

Design of the labwork session introducing the concept of power

In this session, the concept of power is introduced as a 'debit of energy', the quantity of energy transferred per unit of time, in order to differentiate energy (function of state) from power (a physical quantity that quantifies the extent of interaction between systems).

The labwork session includes five activities (see the labwork sheet given in table 2). These different activities are analysed according to our categorisation of knowledge. The theoretical world is subdivided into two sub-categories: the theoretical model with principles, physics concepts and their formal relations, and the numerical model with quantitative relations involving numbers. The measurement apparatus is a sub-category of the world of objects and events.

The first activity requires setting up the experiment, reading measurements, then writing them on a given table. The students mainly have to work in the world of object and events with the possibility of relating the elements

Andrée Tiberghien

of this world with the physical quantities which belong to the world of theory and model.

The second activity obliges students to process the numbers obtained from the measurements mathematically, in order to find a mathematical relation including a coefficient. In our categorisation, this treatment belongs to the numerical model.

The third activity asks students to give a name to the coefficient. This unusual activity is intended to help students to create a meaning for this mathematical coefficient that they have just calculated. To find a name, they can relate this coefficient to the physics concepts of energy and time. They also can relate this coefficient to the objects and events of the experiment. The richest and the most relevant meaning from the physics point of view should involve relations between concepts, measurements and the experiment.

The fourth activity is devoted to enlarging the meaning of power by inserting power into the symbolic representation, the energy chain. Power should be associated to an arrow which implies two systems (at least). For this activity, students can stay in the world of theory and model while establishing relations between several sub-levels, theoretical model concepts, and numerical model.

The fifth activity is also rather rare in typical teaching (Tiberghien et al., 1998). The students have to predict; this prediction rests both on the experiment and the theoretical model. This activity requires establishing relations between the two worlds.

Results

The experiments took place in different classes of the French upper secondary schools as part of regular teaching. However, the four teachers of these classes belonged to a research-development group, so they elaborated a teaching content, and used it in order to discuss it during regular meetings. One or two students' groups from each classroom were video-taped during the labwork session on power. A total of six groups (2 students) were videotaped and their dialogues were fully transcribed (Bécu-Robinault, 1997a).

Results from written reports of all students of the classrooms
The written reports of the 116 pairs of students were analysed. The majority of students gave correct answers (for example, 59% of the answers pro-

Labwork activity and learning physics

We want to heat water with an immersion heater (or similar apparatus) and to make an energy analysis of this situation. From the measurements that you will make, we will ask you to look for a mathematical formula involving energy which will allow you to introduce a new physical quantity.

(Then the list of apparatus is given with the way of transforming Watt x hours into Joules: electrical supply, container with water, immersion heater, thermometer, energy meter (in Watt x hours), chronometer, ammeter, voltmeter.)

Activity N° 1: set up these apparatus in such a way that they will correspond to the energy chain below.

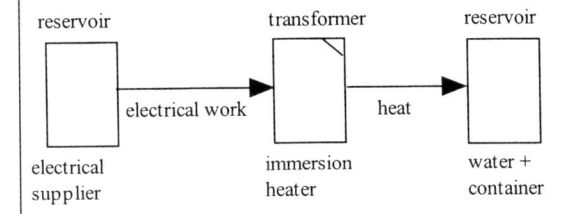

```
reservoir           transformer          reservoir
  ┌──┐                 ┌──┐                ┌──┐
  │  │──────▶──────────│  │──────▶─────────│  │
  └──┘ electrical work └──┘    heat        └──┘
electrical             immersion          water +
supplier               heater             container
```

Making measurements.
　Before starting the measurements, put the disk index in front of the mark.
　Have the chronometer starting when you plug the immersion heater, note the initial temperature t_i and write the results in the table below. Note U =　V and I =　A
　We will not follow the evolution of the temperature. For the final temperature final t_f, plug off the immersion heater when the disk index passes the mark and wait for the stabilisation of the temperature before reading it. (A table is presented).

Activity N° 2 : For this experiment, we wish to be able to predict the results without needing to make all the measurements.
　In order to do that, look for a simple mathematical formula which would allow you to relate the quantity of transferred energy E and the duration of heating t. In this activity, no other physical quantity than E and t is to be taken into account. Let E =f(t) be this formula. Use the units of the international system.
　Write this formula below and briefly indicate how you obtained it.

Activity N° 3: In this relation, you should have introduced a physical quantity which has never been mentioned in this labwork activity. Suggest a name or an expression which corresponds to the meaning of this physical quantity .

Activity N° 4: On which part of the chain (activity 1) would you represent this physical quantity? Put it on this chain!

Activity N° 5: On which elements of the set up that have been represented on the chain does this physical quantity depend?
　What would be necessary to modify on this set up in order to modify the value of this physical quantity? You have the laboratory apparatus available (the teacher puts different apparatus on the table).

Table 2: *Labwork sheet for the quantitative labwork session: introducing of power*

187

Andrée Tiberghien

pose a name which is related to an energy flow rate). They also associate the power to an energy transfer on the energy chain appropriately (77%). Most of them (69%) are able to propose appropriate modifications to the experiment. In general, the students seem to deal with the knowledge as they were expected to do. This figure includes a small number of students who processed knowledge as expected, even though elements of the knowledge being processed were incorrect, even if it is not correct. Therefore the students which process knowledge in an incorrect way but do process it in the perspective that it is requied are counted in the percentage.

Analysis of the dialogues transcriptions of six pairs of students
The videotapes were transcribed and analysed with regard to the modelling categories. These verbal productions were cut into units of analysis. A unit is the minimum set of dialogue turns which has a meaning in itself. Each unit can include one or several modelling levels.

The results are presented by activity in order to compare the intended students' activity and the activity that was effective from the modelling point of view.

Activity 1: Handling and measurement
For all groups this activity was the most time consuming, taking about 60% of the total time for the session (1h30). The results show that even if the students deal with all categories of modelling in both worlds, they spend most of their time making measurements and communicating about objects or events (figure 7). This result is not surprising. However the poor number of relations between the objects and events of the experimental setting, the measurements, and the corresponding physical quantities is noteworthy. These activities of handling and measurement, which are common to most labwork activities, do not seem to lead students to establish relations between objects and events, measurement and physics concepts.

Labwork activity and learning physics

[Chart: Activity 1: Handling and measurement, N = 549. Bars showing "One level only" and "Relations between levels" across Theoretical Model, Numerical Model, Measurement, Object/Event.]

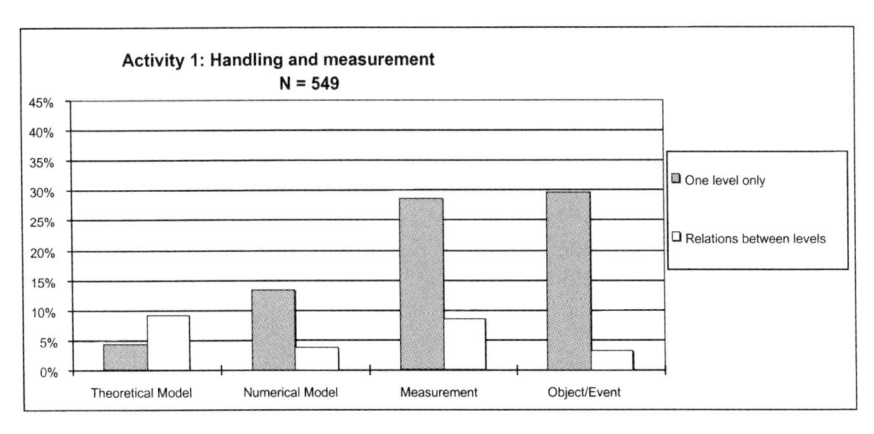

Figure 7: *Levels of modelling and their relations involved in activity 1 (total number of interventions: 549)*

Activity 2: Data processing
Activity 2 is also a very typical labwork activity. The students have to process the data that they just collected by making measurements. As might be expected, the students deal mainly with their numerical data (numerical model) and the corresponding concepts (theoretical model) (figure 8).

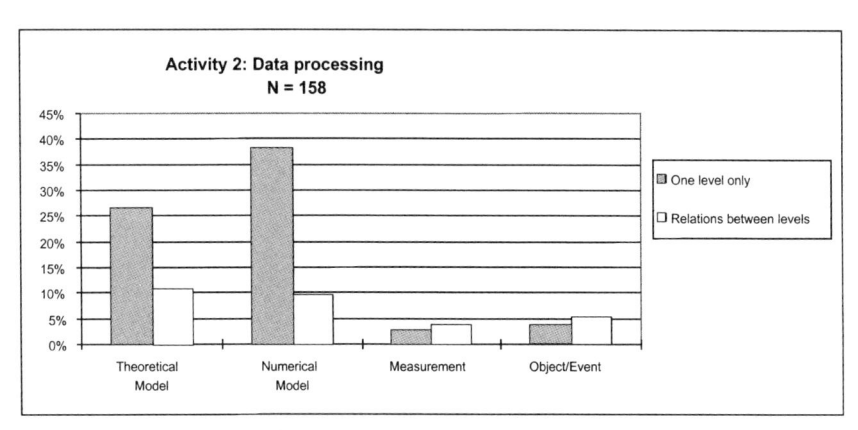

[Chart: Activity 2: Data processing, N = 158. Bars showing "One level only" and "Relations between levels" across Theoretical Model, Numerical Model, Measurement, Object/Event.]

Figure 8: *Levels of modelling and their relations involved in activity 2 (total number of intervention: 158)*

Andrée Tiberghien

Activity 3: Assigning a name
This activity of assigning a name to a quantitative parameter is rather unusual. Contrary to our prediction, students dealt mainly with the theoretical model and to a lesser extent with the numerical one, establishing some relations between them. They did not deal with the objects and events of their experiment (figure 9). According to our learning hypothesis, constructing a meaning for a physics concept requires the construction of a network of relations (Reif & Allen, 1992). This result shows that this network is first established inside the theoretical model. This result ought to be confirmed through other similar studies, however. It is all the more likely since the cognitive cost of establishing links between similar entities is less important than establishing links between very different entities like concepts and material objects that are directly perceived.

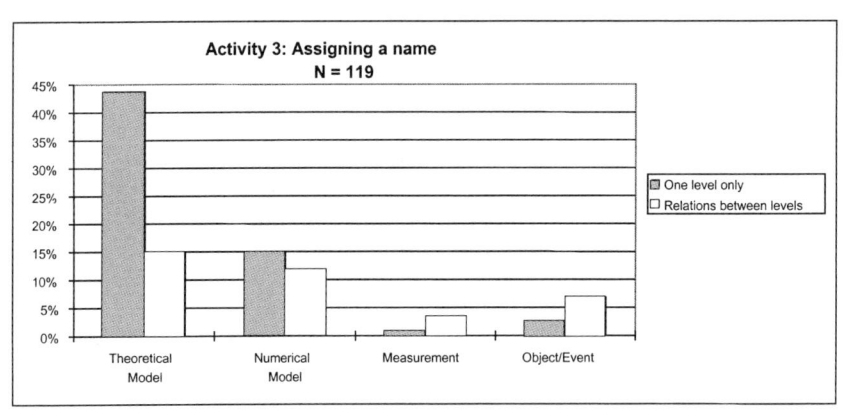

Figure 9: *Levels of modelling and their relations involved in the activity 3 (total number of interventions: 119)*

Activity 4: Inserting power into a symbolic representation
This activity is also atypical. Figure 10 presents a different picture than that of the previous activities, in that relations were made almost exclusively between concepts (theoretical models) and the objects and events of the experimental setting. The students established qualitative relations only; this is favoured by the representation of the energy chain which involves only qualitative aspects. For example a student (J) says: '... *it is in relation to the immersion heater. We have to put it [the coefficient named power] here, because if it [immersion heater] is powerful, it [coefficient] is going to decrease; if it is not powerful, the duration [of heating water] is going to increase ...*'

Labwork activity and learning physics

[Chart: Activity 4: Inserting power into a symbolic representation, N=114, showing bars for Theoretical Model, Numerical Model, Measurement, Object/Event, with "One level only" and "Relations between levels" categories]

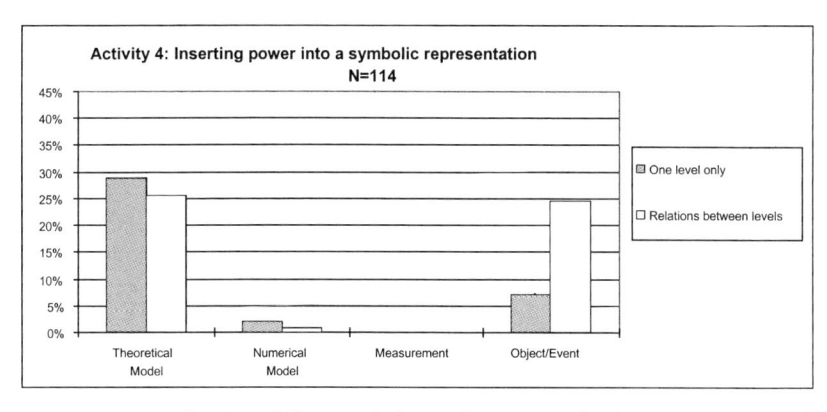

Figure 10: *Levels of modelling and their relations involved in activity 4 (total number of interventions: 114)*

Activity 5: Modifying the value of coefficient (power) by modifying the experiment

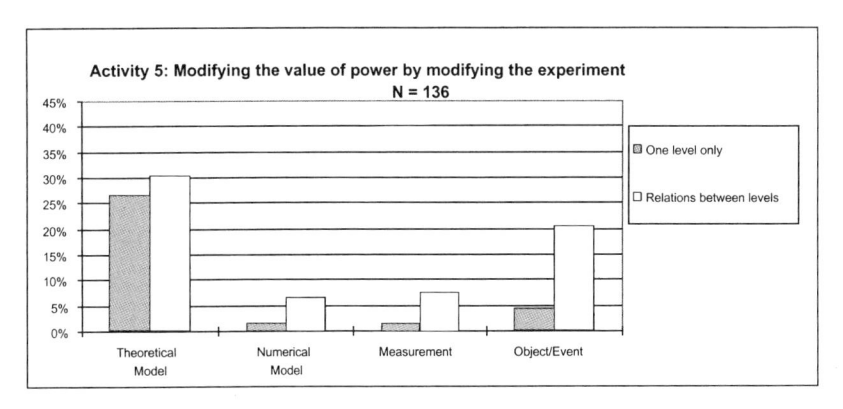

[Chart: Activity 5: Modifying the value of power by modifying the experiment, N = 136, showing bars for Theoretical Model, Numerical Model, Measurement, Object/Event, with "One level only" and "Relations between levels" categories]

Figure 11: *Levels of modelling and their relations involved in activity 5 (total number of interventions: 136)*

This activity is the richest in terms of relations where all levels are involved (figure 11). However the numerical model and measurement were less commonly mentioned than the theoretical model and objects and events. Like in the previous activity, the objects and events are involved mainly in relation to elements of other categories whereas the theoretical model was often used on its own. Again the relations established are mainly qualitative but some involve quantitative aspects.

Andrée Tiberghien

Discussion about quantitative modelling

As shown in figure 12, the successive activities led the students to deal with different aspects of physics knowledge. According to the categorisation presented, the three first activities are similar: the students independently process concepts, numerical values and elements of the experiments. The relations between them are important only in the two last activities. These are the only activities for which the questions require these relations. If the students answer these questions, they cannot avoid establishing such relations. Moreover, a deeper analysis shows that these relations are mainly qualitative. Therefore, even if the students spend most of their time dealing with numbers, particularly during the measurements and data processing, when they reason about the experiment in terms of physics concepts, they do so qualitatively. Quantitative aspects are involved when the students process their data, and when they extend this processing they do so towards physics concepts rather than towards the experiment.

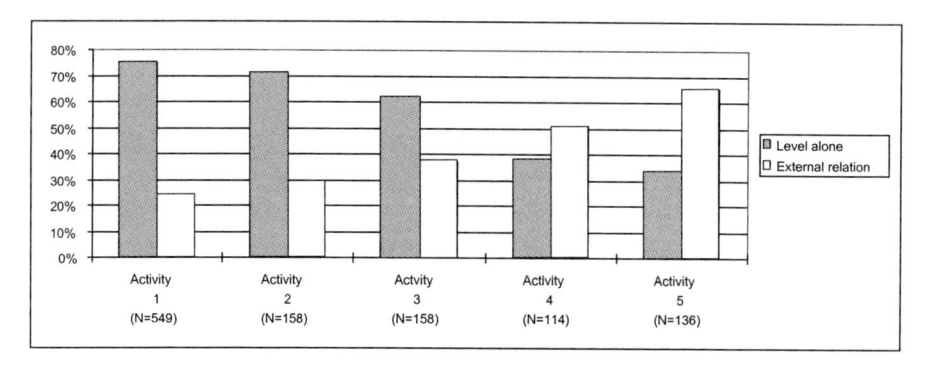

Figure 12: *Levels of modelling used alone or in relations during the labwork*

Discussion and conclusion

This categorisation of knowledge is a way to discriminate between different activities performed by students during labwork. First, it shows the important influence of the questions asked in terms of promoting particular cognitive activities. If the aim is that students construct relations between physics concepts and the experiment, is not enough that students are involved in handling, measurement and data processing. The questions seem to influence students' cognitive activities very strongly.

Secondly, with this categorisation of knowledge it is possible to make some hypotheses about how students learn different aspects of physics concepts. The first labwork session, dealing exclusively with qualitative aspects, shows that relations between the two worlds require a development of knowledge about both, even if the experiment is very familiar. A reinterpretation of experiments, to see them in another way than the spontaneous one in order that they fit with the theoretical approach, is necessary. Our results lead us to think that, during labwork, a specific activity should be designed to allow this reinterpretation of the experiments whereas usually it is implicit or considered as obvious. The second labwork session, dealing with the introduction of the power concept, shows that among the numerous possible relations, students favour two types. The first type of relation is internal to the theoretical world. This type of relation occurs between the facets of the concepts made explicit within natural language, or within mathematical formal language, or within numerical relations coming from measurements. The second type corresponds to relations between concepts and elements of the experiment which are established in a qualitative form. A third type of relation, involving elements of the experiment and quantitative aspects of concepts, is very rare. It may have too high a cognitive cost to be possible the first time that students encounter a physics concept.

It appears that even if labwork sessions include qualitative tasks, they can offer a large variety of activities if they are carefully designed. It also appears that quantitative tasks do not directly lead students to establish relations between on one hand quantitative relations of physical quantities, and on the other hand the material elements of the experiment. Quantitative relations seem to be restricted to a treatment at the conceptual level, when students first encounter a concept. The design of labwork sessions should specifically foster quantitative aspects at the level of physics concepts, links between qualitative and quantitative aspects, and their relations with the experiment. Each task should not demand too many different cognitive activities at the same time because students only use a few. However, these hypotheses have to be tested more systematically across a range of different types of labwork.

Andrée Tiberghien

References

Bécu-Robinault, K. (1997a). *Rôle de l'expérience en classe de physique dans l'acquisition des connaissances sur les phénomènes énergétiques*. Thèse de doctorat, Lyon: Université Claude Bernard, Lyon 1.

Bécu-Robinault, K. (1997b). Activités de modélisation des élèves en situation de travaux pratiques traditionnels : introduction expérimentale du concept de puissance. *Didaskalia*, 11, 7-37.

Bunge M. (1973). *Method and matter*. Derdrecht-Holland: D. Dcidel publishing company.

Lunetta, V.N. (1998). The school science laboratory: Historical perspectives and contexts for contemporary teaching. In K. Tobin & B. Fraser (Eds) *International handbook of science education*. Kluwer, The Netherlands.

Rozier, S. (1988) *Le raisonnement linéaire causal en thermodynamique élémentaire*, Thèse de physique. Paris: Université Paris 7.

Shipstone, D.M. (1985) 'On children's use of conceptual models in reasoning about current electricity', In R. Duit, W. Jung & C. v. Rhöneck (Eds.), *Aspects of understanding electricity. Proceeding of an international workshop* (pp. 73-93). Ludwigsburg: IPN, Kiel.

Tiberghien A. & Megalakaki O. (1995). Characterisation of a modelling activity case of a first qualitative approach of energy concept. *European Journal of Psychology of Education*, 10, (4), 369-383.

Tiberghien, A. (1996). Construction of Prototypical Situations in Teaching the Concept of Energy. In G. Welford, J. Osborne, P. Scott (Eds.) *Research in Science Education in Europe*. (p. 100 – 114). London: Falmer Press.

Tiberghien, A., Veillard, L., Le Maréchal, J.F., and Buty, C. (1998). Analysis of labwork sheets used in regular labwork at the upper secondary school and the first years of University. Working paper n°3 European Project: Labwork in Science Education Lyon: Université Lumière Lyon 2.

Reif, F & Allen, S. (1992). Cognition for interpreting scientific concepts: A study of acceleration. *Cognition and Instruction*, 9, 1-44.

Welzel,M., Haller, K., Bandiera, M., Hammelev, D., Koumaras, P., Niedderer, H., Paulsen, A., Bécu- Robinault, K., and von Aufschnaiter, S. (1998). Teachers' objectives for labwork. Research tool and cross country results. Working paper n°4 European Project: Labwork in Science Education Lyon: Université Lumière Lyon 2.

Modelling student's cognitive activity during the resolution of problems based on experimental facts in chemical education.

Jean-François Le Maréchal

Problem solving in chemical education is mostly based on experimental facts, belonging to our real world, that must be 'translated' into chemical substances, their symbolism, chemical equations, etc., that we call the chemists' reconstructed world. These two worlds have similar structures which include: a sub-domain of objects, of representations, of events, of properties and of rules. This categorisation allows an interesting view of the chemical cognitive activity of students. Students' difficulties arose most frequently when they had to relate either two sub-domains of the reconstructed world, or equivalent sub-domains of the two worlds.

1. Introduction

The difficulties students have in acquiring conceptual knowledge in chemistry has been well documented in the case of the particulate nature of matter (Driver, 1983; Novick & Nussbaum, 1981). Nevertheless, attempts to describe students' mental processes while acquiring complex knowledge, such as the concepts involved in physics or in chemistry, are rare (Tiberghien, 1994; Vosniadou, 1994).

Although the questions used in this research are very simple for chemists (the text of these questions is shown below), they cannot be solved by using a classical algorithm such as the use of stoichiometry (Frank, Baker & Herron, 1987). Instead, they require a molecular vision of the situation and the aim of this paper is to find a way to describe the molecular approach to problems. The questions proposed to our students involved them in a problem solving situation, defined by Hayes (1980) as a question to which an immediate answer cannot be given, although no mathematical steps are necessary to solve them (Gendell, 1987).

The aim of this work is to propose a model which can account for the cognitive activity of chemistry students as they attempt to solve simple problems.

2. Specific features of chemistry

Many chemistry tasks proposed to students during labwork may be solved at two levels of interpretation: a level that does not involve a molecular view of the system and a level that is deeply anchored in a molecular description. For example, work on acidity can involve simple concepts, such as the colour of pH-paper, or more sophisticated ones, such as the potential of a membrane electrode. Both of these avoid any molecular description. The same activity, however, may also be described in terms of H+ ions or might involve the strength of chemical bonds in weak acids. These are examples of molecular descriptions.

Professional chemists are experienced in solving their problems at this level. For them, molecules are real objects with properties, a shape and motion, which change during chemical reactions.

Pupils at an early stage of their education in chemistry ignore most, if not all, of this invisible molecular world and have to build this knowledge. They must build the meaning of the concepts related to the molecules, and the rules used for reasoning about them from the simple experiments they perform during labwork sessions, from teachers' explanations, and from solving problems. The aim of the work described in this chapter is to try to describe pupils' mental activity at this level of detail, involving features which are specific to chemistry.

3. The two world mental organisation

I have chosen to describe the whole molecular world as a reconstruction of the real world, with the same kind of structural organisation. I know of no studies that apply this idea to chemical education, though an approach of this kind has been successfully used in physics education (Tiberghien & Megalagaki, 1995 ; Quintana-Robles, 1997).

The real world everyone lives in can be described in terms of objects (chairs, cars...), of events (the teacher closes the door), and of properties (the table is white). Objects can be described in different ways: with words, with drawings. Finally, our life is regulated by rules that are socially accepted, such as 'stop when the traffic light is red', or based on empirical evidence: 'a sharpened knife cuts better'.

It might have been possible to choose a different structure for the real world, but this one seemed to fit nicely when transposed into our reconstructed world. Later I will explain more fully what I mean by a 'reconstructed world', made of reconstructed objects (molecules), of reconstructed events (their transformations), and so on.

4 . The structure of the two worlds

I will now identify, for each of the two worlds, real and reconstructed, five sub-domains by using them to describe a classical chemical situation: an acid-base titration. When a non-chemist and a chemist are faced with this same chemical experiment, both will see the same things and the same events, but the chemist will be able also to have a molecular vision of the experiment. I will present a blow-by-blow account of this vision in term of the concept of a reconstructed world.

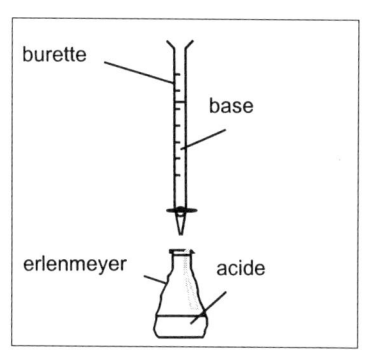

Figure 1. *In this titration, glassware consists of a burette and an erlenmeyer, the reagents are an aqueous solution of sodium hydroxide in the burette and, an aqueous solution of ethanoic acid with few drops of phenolphthalein in the erlenmeyer*

4.1. The sub-domain of objects

In the real world, in the sub-domain of objects, there is glassware and two colourless liquids. The colourless liquids are real objects. The reconstructed objects corresponding to these liquids are ethanoic acid and sodium hydroxide. For non-chemists, ethanoic acid is just a word that can be written on the label of a flask of the colourless liquid but, for chemists, this word, when given to a substance, calls to mind many concepts and allows them to differentiate real objects (liquids) from reconstructed ones (chemical substances).

4.2. The sub-domain of representation

For non-chemists, ethanoic acid can be represented by words (colourless liquid, the liquid that has been poured into the erlenmeyer flask...) or by a drawing. In figure 1, there is no need to be a chemist to understand that there is a liquid in the erlenmeyer flask. All these semiotic forms belong to the sub-domain of the representations of the real world.

Representations of the real world are also useful for chemists. Chemists can also work with reconstructed representations such as CH_3COOH (or any of the other ways to express this chemical structure). Chemical symbols, developed formulae, and crystal representations with hard sphere drawings are other reconstructed representations of reconstructed objects.

4.3. The sub-domain of events

During a titration, several events occur, such as the base dripping from the burette, but one event is of special interest: the colourless liquid in the erlenmeyer flask turns pink after a while. This event is unexpected for non-chemists. This is a real event which everyone will agree about.

For chemists, this real event indicates that the titration is over: they consider the titration is represented by the chemical equation:

$$CH_3COOH + Na^+ + OH^- \rightarrow CH_3COO^+ + Na^+ + H_2O$$

This equation is a reconstructed event. Chemists associate the real events (colour change) with reconstructed ones (chemical equation).

4.4. The sub-domain of properties

For non-chemists, the liquids involved in this titration have several properties. For example, the solution in the erlenmeyer flasks tastes acidic (it is, in fact, the acidity of vinegar). We can say that this is a real property. The Romans knew the property of this beverage even though they had no chemical ideas in mind.

For chemists, the acidic property of ethanoic acid solutions is more than just a taste. It relates to a theory, for example the Bronsted theory of acidity. I will call this Bronsted acidity a reconstructed property of the solution.

4.5. The sub-domain of rules, models, laws and theories

We saw earlier that real rules could be divided into social and empirical ones. In the case of titrations, we could imagine that any non-chemist could follow a protocol that reads :

> Take 10 cm^3 of the liquid called 'acid' and add a few drops of the liquid in the bottle labelled 'indicator'. Add the liquid in the burette drop by drop until the solution turns pink. Note the volume delivered by the burette.

The rule 'stop adding the base when it turns pink' is an empirical rule that needs no chemical competence. As soon as someone is taught elementary manipulation of the glassware, s/he may follow the titration instructions above and obtain excellent data. So I will consider this empirical rule to be part of the real world.

This empirical rule is associated with a theoretical background about acidity: the role of the acid-base indicator, the pH jump during the titration and so on. These correspond to reconstructed rules. Theories, laws, models, possibly re-written as mathematical formulae or rules can be considered to belong to a sub-domain of the reconstructed world. In the example above, the change of colour during the titration is unexpected to non-chemists because they have no access to the reconstructed rules. It is the theories, laws and models which provide interpretations and predictions for chemists.

4.6. Representation of these sub-domains

To represent the cognitive activity of pupils, it is convenient to represent these two worlds and their five sub-domains graphically as in figure 2.

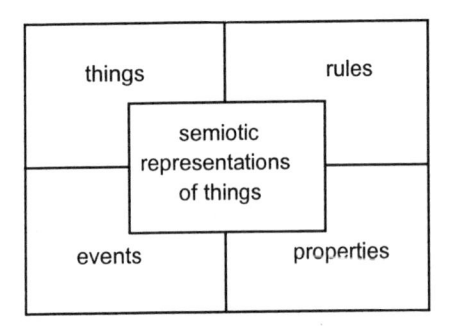

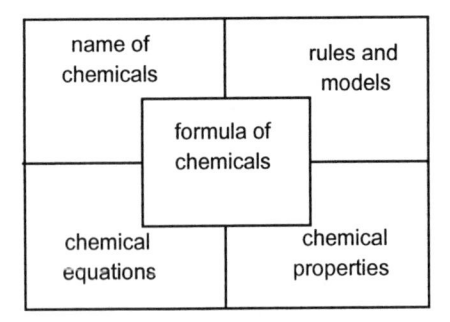

Figure 2 : The real (left) and the reconstructed (right) worlds and their 5 sub-domains.

5. Use of the model of the two worlds

This categorisation gives us an interesting view of chemical activities. Students' difficulties happened most frequently when they had to relate either two sub-domains of the reconstructed world, or equivalent sub-domains of the two worlds. In this paper, I will present several research studies carried out in different schools. All the students involved were 16-17 years old and were in the first form of upper secondary school in Lyon, France in 1997. The first study is based on a test given to 96 pupils. The second study was part of a labwork exercise that was given to 17 pupils. The last study involved a set of four tests given to 84 students.

5.1. Test 1

Test 1 started with the following questions :
1- A sodium ion is Na^+
   A chloride ion is Cl^-
   a) What is the formula of sodium chloride?
   b) Which of these ions is the anion and which is the cation?
2- a) How would you prepare a sodium chloride solution?

The responses to questions 1a and 1b confirmed the pupils had learned about ions and ionic solutions during the previous year: in 1a, there were 87 correct answers (91%) and in 1b, 72 correct answers (75%). Nevertheless, these pupils had probably never prepared any solution for themselves during labwork, as in most labwork sessions, all solutions are ready on the bench.

A large number (see table 1) of pupils wrote that they would prepare sodium chloride solution by mixing sodium ions and chloride ions. The high frequency of this wrong answer seems to indicate a persistent error as, even in the next form, many pupils still make the same mistake. Just as conceptions may reflect coherent reasoning by the pupils, though different from the scientifically accepted version (Johsua & Dupin, 1993), these mistakes may make sense within the pupils' view of ionic solutions.

N = 96 3 different classes in 3 different schools	correct answers	Salt + water	10 (10%)
	main wrong answers	mix Na^+ and Cl^-	46 (48%)
	other wrong answers	HCl + Na HCl + salted water...	6 (6%)
	no answer		34 (36%)

Table 1: *answers of test 1 grouped into categories. The 10% of correct answers correspond approximately to 3 pupils per class.*

We could interpret this observation by proposing that pupils hold alternative conceptions of ionic solutions, but we shall see later that other pupil mistakes would require us to propose other alternative conceptions. Instead of trying to interpret these observations in terms of known alternative conceptions (Tiberghien, Jossem & Barojas 1998), I will propose an interpretation based on the two-world model.

Figure 3 shows the concepts used in table 2 to account for pupils' interpretations.

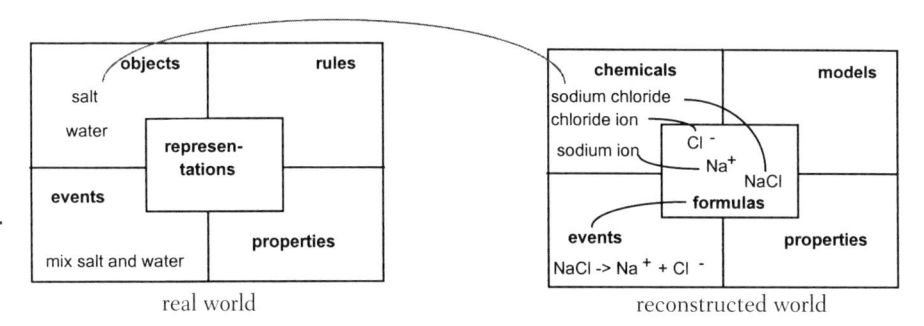

objects	rules		chemicals	models
salt water	represen- tations		sodium chloride chloride ion sodium ion	Cl^- Na^+ NaCl formulas
events mix salt and water	properties		events NaCl -> Na^+ + Cl^-	properties
real world			reconstructed world	

Figure 3: *use of the two-world model to interpret the pupils' wrong answers about the preparation of a sodium chloride solution : 'mix sodium ions and chloride ions'.*
note: *according to 'pedagogical fashions' the NaCl $\rightarrow$ Na^+ + Cl^- equation representing the dissolution of salt may have other formulations.*

Jean-François Le Maréchal

concepts	sub-domains
sodium chloride	as a reconstructed object (see note)
sodium ion	reconstructed object
chloride ion	reconstructed object
the preparation of a sodium chloride solution	real event

Table 2: *Concepts and their corresponding sub-domains to account for pupils' interpretations of test 1 answers.*
note: *sodium chloride can be either considered as a real object (common salt) or as a reconstructed object (the NaCl substance). In this case, as the question of the test had just involved the pupils in the formula "NaCl", they are very likely to have the reconstructed object in mind.*

Solid lines are used to link concepts that are implicitly or explicitly connected by pupils. In this model we can see that the question: 'how would you prepare a sodium chloride solution' is an event of the real world. As the concepts pupils have immediately available are in the reconstructed world, there are no straight connections. Pupils need to establish a connection between 'sodium chloride' and 'salt' (the dotted line in figure 3). We can understand the internal coherence of pupils' reasoning by assuming that it is cognitively less expensive to remain in the same world, so pupils connect a reconstructed representation and a reconstructed event (see figure 3) and give an incorrect 'translation' of the reconstructed event: mix Na^+ and Cl^-.

It is surprising to observe that, even though pupils do not know the reconstructed world, our model shows that everything occurs as if they were reasoning in this world, to work out a solution.

5.2. Labwork on acidity

The labsheet for a labwork task given two months after the beginning of the year read as follows :

pH measurements
(a) With a pH meter, measure the pH of ordinary tap water.
(b) Add 2 drops of a sodium chloride solution to the tap water and measure the pH again.
(c) Add to the same beaker 2 drops of a solution of hydrochloric acid and measure the pH again.

solution	chemical composition of the solutions	measured pH values
(a) = tap water		
(b) = (a) + sodium chloride solution		
(c) = (b) + hydrochloric acid		

Table 3: *table provided in pupils' labsheet to help them organising their experimental data.*

Summarise all the values in table 3 :
Question: *From the experiments (a), (b) and (c), find what chemical species is responsible for the acidity of a solution.*

According to the official curriculum, these pupils have learned the qualitative basis of the pH concept during the previous year. They were able, for instance, to answer confidently that pH can be measured either with litmus paper or with a pH-meter.

Expected answers

I expected the following reasoning scheme, and the table provided to pupils (table 3) should have guided them in this way:

(a) In tap water, there is only H_2O, and the pH is approximately 7 (this works with French tap water, but not with distilled water, always slightly acidic).
(b) When sodium chloride is added, the species in solution are H_2O, Na^+ and Cl^- and the pH is still around 7.
(c) When hydrochloric acid is added, the species in solution are H_2O, Na^+, Cl^- and H^+, and the pH is about 3.

The only newly added species, when the pH dropped to 3, is H^+. Therefore, H+ is the chemical species responsible for the acidity of a solution.

This question about acidity led to a surprising number of wrong answers: 7 (41%) of pupils answered: 'The Cl^- ion is responsible for the acidity of the solution', 1 (6%) answered correctly and 9 (53%) gave no answer. Some of the pupils who answered that Cl^- was responsible provided the following

203

Jean-François Le Maréchal

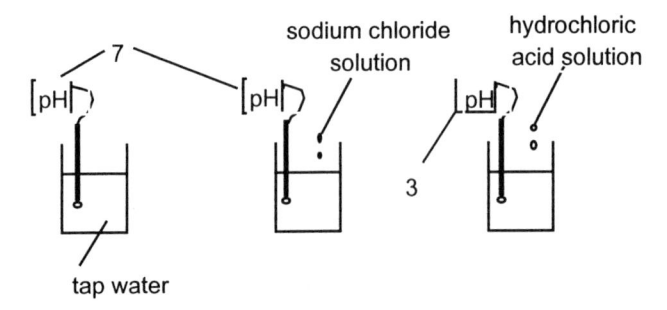

Figure 4: *this scheme graphically summarised the three steps of the experiment realised by pupils.*

explanation: 'because Cl⁻ ions are also present in step (b)'. Even though this question is about a different chemical topic from the preparation of ionic solutions, the same two-world model can be used to interpret it. As with the previous interpretation (figure 3), figure 5 shows the concepts needed to answer the question. The solid lines show the connections easily available to pupils. I have also connected pairs of sub-domains when the term-to-term connections of all their elements are straightforward, for the sake of clarity. Pupils had all the connections from the real events (the pH measurement) to the reconstructed properties at the core of the question, apart from one – shown by the dotted line.

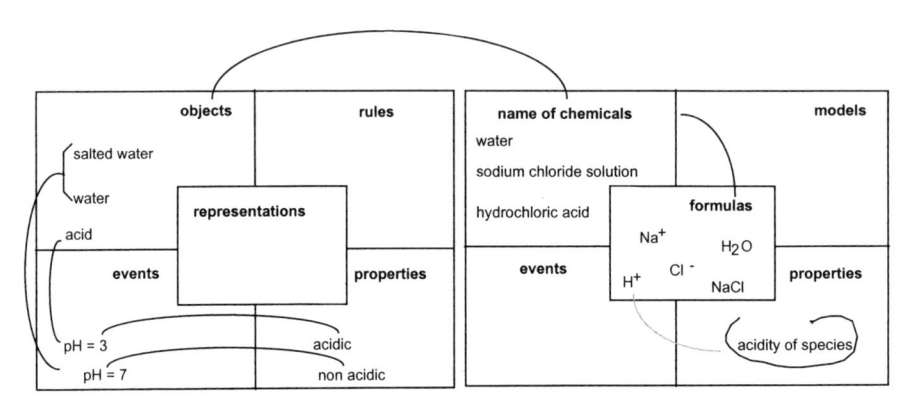

Figure 5 : *use of the two world model to interpret the pupils' answers about the species responsible for acidity in solution.*

Modelling student's cognitive activity

We can therefore conclude that this connection is a difficult one. This implies that the fact that a species (a reconstructed object) may have a property (the reconstructed property of acidity) is not appreciated by students who are not told explicitly. When teachers claim that Cu^{2+} has a blue colour, that OH^- is a base, etc., they implicitly formulate the hypothesis that a species may have a property. According to our two-world model, this is a way of reasoning in the reconstructed world which has to be taught.

5.3. Controlling the connections between sub-domains

Now that we are able to model the cognitive processes involved in pupils' understanding of the molecular aspect of chemistry, we can see if we can influence pupils' reasoning so as to increase their use of certain connections in our two-world model, or decrease others.

5.3.1. Methodology

The third experimental part of this work involved 84 pupils, who had completed the first term of the first year of upper secondary school. This new group of pupils had learned more about aqueous ionic solutions, as their curriculum made them work theoretically and practically on this subject. So we divided them into four groups (about 20 pupils each); the first group were given test 1 as above while the second, third and fourth groups were given slightly different tests (tests 2, 3 and 4 respectively).

Text of test 2
Cadmium ion is Cd^{2+}
Benzoate ion is $C_7H_5O_2^-$
What is the formula of cadmium benzoate?
How could you prepare a cadmium benzoate solution?

Text of test 3
Cadmium benzoate is a white powder. How could you prepare a cadmium benzoate solution?

Text of test 4
How could you prepare a cadmium benzoate solution?

5.3.2. Analysis of the tests

At this school level, most pupils will never have heard of cadmium or benzoate ions. This is confirmed by the results from tests 3 and 4 where the formulae were not given, and no students used these terms. In the terms of our two-world model, the words 'cadmium' and 'benzoate' are reconstructed objects, while Cd^{2+} and $C_7H_5O_2^-$ are reconstructed representations. This was not necessarily the case with sodium chloride as the word 'salt' exists, is known to the pupils, and we cannot know how many of them associate sodium chloride and salt during the test.

Group 2 pupils had a test set completely in the reconstructed world, with a link between two of its sub-domains: chemicals (cadmium ion) and formulas (Cd^{2+}).

Group 4 pupils are also asked a question in the reconstructed world, but only in the sub-domain of the reconstructed objects. There are no links with any other sub-domains of the reconstructed world.

Group 3, on the other hand, had a text which made a link between the two worlds, as the sentence '*Cadmium benzoate is a white powder*' connects a reconstructed object (cadmium benzoate) with a real object (white powder).

From the analysis above, we might expect :

– several good answers to test 3, as its text provides a connection between the two worlds,
– several wrong answers to test 2 as it induces a non useful connection in the reconstructed world,
– a lot of 'no responses' to test 4, as the text makes no links and it may not incite the pupils to create any, or else results similar to those of group 2. This second outcome would suggest that it is enough to be in a world for the pupils to stay in that world and create links by themselves, even without knowing much about this reconstructed world.

5.3.3. Results

We analysed the pupils' answers using the following categories:
– A1- The correct answers: 'put *it* into water'. The *it* was the powder or the cadmium benzoate.
– A2- The expected wrong answers, where pupils proposed mixing the ions, identified either by their names or their symbols.
– A3- Other wrong answers.
– A4- No answers.

The results are summarised in table 4.

N = 84 3 different classes random distribution of tests		test 1 (N=20) NaCl	test 2 (N=21) $C_7H_5O_2^{2-}$ + Cd^{2+}	test 3 (N=20) white powder	test 4 (N=24) benzoate cadmium
A1	put it into water	0	0	7 (35%)	0
A2	mix ions	18 (90%)	13 (62%)	11 (55%)	13 (56%)
A3	other	1 (5%)	4 (19%)	1 (5%)	5 (22%)
A4	no answer	1 (5%)	4 (19%)	1 (5%)	5 (22%)

Table 4: *answers of four groups of pupils to the tests 1-4.*

5.3.4. Discussion

Pupils who answered test 1 proved that teaching aqueous ionic solutions for a term does not greatly change the deep way that students see this topic. From a methodological point of view, the fact that the results of the tests given before and after can be compared is reassuring.

Test 2 is an exact copy of test 1 except that the familiar term 'sodium chloride' has been replaced by the unknown term 'cadmium benzoate'. The results can be quantitatively compared. There are no right answers that we might interpret by a link between the real and the reconstructed worlds. An interesting difference between tests 1 and 2 comes from the answers to A3.

The results of test 3 are as predicted. This is the only group of pupils with an appreciable number of right answers: 35%. These pupils did establish the connection between both worlds. Comparing this with the results of tests 2 and 4, the number of 'other' responses or 'no response' is lower, just as low as the results of the test with sodium chloride. The real object 'white powder' made a connection with the reconstructed object 'cadmium benzoate' which created a sense of familiarity with this esoteric substance.

Last, the *a priori* analysis of test 4 raised two possibilities: either a lot of null responses or results similar to group 2. The second alternative seems to be borne out; the results of groups 2 and 4 are reasonably similar. This suggests that pupils can create links in the reconstructed world even when we do not provide any. The fact that pupils create links inside the reconstructed world supports the idea that they try to work in the reconstructed world even though they know very little about it.

Most of our expectations about tests 1, 2, 3 and 4 are fulfilled. Once more, the two-world model proves to be an efficient tool for understanding the reasoning of pupils in chemistry. In addition, it seems possible to increase or decrease the likelihood that pupils will create links between the two worlds. this result has implications for the writing of chemistry exams. The two-world model provides a way to grade the difficulty of problems: by suggesting links, for instance by adding apparently insignificant comments such as 'a white powder', you may alter the cognitive activity of pupils.

6. Conclusion

This paper has proposed the seed of a theory that can explain the cognitive activity of pupils learning chemistry. In fact this model has bee used to interpret other observations in addition to those discussed here: it can be applied widely. The model is simple to use and may be useful for *a priori* evaluation of assessment tasks or documents written for students. The results show that pupils may work in the reconstructed world even though they have not been taught, and as a result they propose (sometimes wrong) solutions to problems or predictions in new situations.

References

Driver, R., Guesne, E. and Tiberghien, A. (Editors) (1985) Children's ideas in science. Milton Keynes: Open University Press.

Frank, D.V., Baker, C.A., & Herron, J.D. (1987). Should Students Always Use Algorithms to Solve Problems? *Journal of Chemical Education*, 64, 514-515.

Gendell, J. (1987). The Solution is Not the Problem. *Journal of Chemical Education*, 64, 523-524.

Hayes, J. R. (1980). The Complete Problem Solver. Philadelphia: Franklin Institute.

Johsa, S., & Dupin, J.J. (1993). Introduction à la didactique des sciences et des mathématématiques [Introduction to the didactics of science and mathematics] (pp 121-192). Paris: Presses Universitaires de France.

Novick.S., & Nussbaum, (1981). Pupils' understanding of the particulate nature of matter: a cross age study. *Science Education*, 65, 187-196.

Quintana-Robles M. (1997). Étude didactique de films comme aide pour l'enseignement de la physique: cas de l'expansion des gaz. [Didactic investigation of the use of films as an aid to the teaching of physics: the case of gas expansion] PhD Thesis. University of Lyon I, France.

Tiberghien, A., & Megalagaki, O. (1995). Caracterisation of a Modelling Activity for a First Qualitative Approach to the Concept of Energy. *European Journal of Psychology of Education.*, 10, 369-383.

Tiberghien, A. (1994). Modelling as a Basis for Analyzing Teaching-Learning Situations. *Learning and instruction.*, 4, 71-88.

Tiberghien, A., Jossem, E.L., & Barojas, J. (1998). Connecting Research in Physics Education with Teacher Education. *ICPE publication*, available at: http://www.physics.ohio-state.edu/~jossem/ICPE/books.htlm

Vosniadou, S. (1994). Capturing and Modelling the Process of Conceptual Change. *Learning and instruction*, 4, 45-70.

A challenge for lifelong science understanding
The role of "lab work" in primary school science

M. Gagliardi, N. Grimellini Tomasini, B. Pecori

This paper describes a teaching experiment carried out with a class of 5th grade pupils (age 9-10). The aim of the experiment was to foster in the students a way of looking at the description of motion in terms of concepts close to the scientific ones and, in particular, in terms of speed. In the discussion we emphasise the role played by the learning environment implemented in the class as a "laboratory of facts and ideas", specially designed to foster the process of construction of ways of looking nearer to those peculiar to disciplinary knowledge and to help the pupils to become aware of the development of their ideas. Activities of this kind can become the basis for the development of a positive attitude toward disciplinary knowledge in the students and can also heavily influence science learning at higher levels of education. The study we shall discuss in this paper was based on the idea that the foundations of secondary science teaching should, and ought to be laid in the primary school by taking advantage of the open-mindedness and curiosity of young children and by letting them appreciate the process of disciplinary content knowledge construction as something valuable and enjoyable.

It is well known that science, and physics in particular, is not the favourite school subject for the vast majority of secondary school pupils. As they move from primary to secondary school, experiencing the teaching of the science subjects, their attitude toward science teaching appears gradually to develop into a negative one.

We believe that primary school science can make a major contribution to modifying this tendency by acting at three levels: by fostering the pupils' construction of new ideas closer to scientific ones, by encouraging them to become aware of their processes of construction, and by helping them to gain personal satisfaction from it.

To do so a "space" is needed where their natural curiosity and receptiveness can be guided to construct new ways of looking which are not spontaneous but can be recognised by them as sensible and powerful instruments for understanding and acting on the reality around them, on the basis of their everyday life experience and of their knowledge about it. We shall call this space a "laboratory of facts and ideas".

"Laboratory" because it has to be perceived as a space where "things" are analysed, constructed and modified. "Of facts" because it should be clear that science has to do with facts and events of the natural and technological world that is with objects that are not only in our mind. "Of ideas" because we think it essential to emphasise that science is a social process by which ideas about the natural world are constructed, discussed and modified continuously and because "everyday concepts mediate the acquisition of scientific concepts" (Scott, 1996, p.327). It is the interplay of facts and ideas that makes science an intellectual adventure worth the effort and the individual construction of knowledge a meaningful process (Arcà et al., 1984; Grimellini et al., 1990; 1992).

The aim of this laboratory is therefore to help the pupils not only to master pieces of scientific knowledge and develop their abilities related to it, but also to gain insights about the processes peculiar to scientific knowledge and about its difference from those peculiar to common-sense knowledge (Scott, 1996).

In this contribution we shall give examples of work in the "laboratory" focusing in particular on the "ideas" component, which appears to be too little developed at the junior levels of science education. Indeed the "laboratory of ideas" is what makes it possible to move from facts based on everyday experience and common-sense knowledge, to facts as results of designed experiments and more specific disciplinary knowledge. By allowing the pupils to recognize as "problems" features of the phenomena that they would not spontaneously recognize, and by giving them the chance to construct new ways of looking that enable them to tackle and solve those problems, the lab work allows the pupils to master the facts more efficiently and rigorously than could be done with their previous knowledge, and makes them aware of the development of their knowledge.

M. Gagliardi, N. Grimellini Tomasini, B. Pecori

A study about motion

In the study described in this paper a "laboratory" about motion was set up in a 5th grade class of 17 pupils (9 females, 8 males), aged 9 to 10, of a primary school in a small town near Bologna.

Motion was chosen because the ideas of time, speed, etc. naturally develop in children at the primary school age level (Piaget, 1946). This natural development can be exploited in science teaching by positively interacting with spontaneous development and shaping it so as to make knowledge construction in science more natural and easier for the children.

The pupils had done no previous school work on the concepts involved in the study nor on other physics topics.

The interaction with the class lasted about 36 hours (divided into 12 sessions) over a two month period.

All sessions were audiotaped and transcribed. Data about pupils' individual learning processes have also been obtained by collecting their solutions to problems set in the classroom, their homeworks, their answers to a questionnaires and by interviewing them. All data were used both to adapt the teaching path on line with the classroom activities and later to evaluate the efficacy of the choices at the basis of the teaching experiment.

The teaching experiment focused on the kinematic description of translational motion and aimed at introducing the following concepts:

- position (frame of reference)
- trajectory and its representation
- distance travelled and corresponding time
- speed in uniform motion

The study aimed also at testing the possibility of making the pupils aware of the differences between the description of motion in physics, and their spontaneous ones.

For this purpose the following interrelated aspects of the process of disciplinary knowledge construction were emphasized:

- examining the phenomenon in an everyday context, with all the complexity peculiar to it, and identifying criteria for performing a schematisation which enables the pupils to construct new instruments starting from the design and analysis of simpler situations;

A challenge for lifelong science understanding

- taking into account that the descriptions to be constructed should be agreed upon and be able to be communicated to others;
- constructing adequate conceptual instruments for a schematic description of the phenomenon by identifying the variables to describe it and the formal language to express the relationships among them which can describe any real situation within the limits of the schematisation;
- operationally defining the quantities involved in the descriptions.
- checking that the conceptual instruments are adequate to describe reality by testing predictions with experimental data.

Lab work about motion

The classroom activities started with the investigation of the motion of a small remote-controlled car and was guided by the task, proposed by the adult, of describing its motion to another pupil who was out of the classroom during the event, so that he could repeat it.

The first part of the classroom work (see figure 1) was then devoted to the construction and the checking of the validity of instruments (frame of reference, co-ordinates, scale drawing, distance travelled) for a non-ambiguous description of the trajectory of the car, which was the first feature identified by the pupils as peculiar to the car motion.

The second part of the classroom activities was devoted to the identification of other instruments necessary for a non-ambiguous description of motion (the concept of speed and its relationship with time and distance), following the identification of speed as another feature of motion necessary for its description. We will discuss a few excerpts of lab work specially suitable for showing the evolution of the interplay of children's ideas, briefly outlining the main steps in this part of the learning path.

The "facts" from which the second part of lab work started were the children runnning along the corridor. The schematisation process, guided by the idea of describing the children's running performance, led to discarding some aspects of the runs, like the footstep length and frequency, and to identifying two variables (time and the previously defined distance) as suitable quantities for comparing the children's performance.

For the purpose of consolidating the pupils' spontaneous strategy by making it explicit and agreed upon, the children were then invited to imagine ex-

Geometrical aspects of the description of motion:
- observing situations of motion (motion of a remote-controlled car) and discussing the problem of its description (comparison of spontaneous descriptions)
- constructing instruments for a non ambiguous description of trajectory (maps, co-ordinates and scales)
- introducing a first quantity specific to motion (distance travelled)
- comparing the measurement of a real distance with the value obtained from the representation of the trajectory, and discussing the problems of uncertainties in measurements

The relationship between d, t and v:
- observing examples of motion (children running), discussing how to describe those situations, and identifying the need for two more variables: time and speed
- constructing and comparing examples of races in couples, based on the intuitive ideas of speed and uniform motion, discussion of the procedures to compare the motion of bodies moving at different speeds
- looking for the mathematical relationships among d, t and v that allow the previously analysed situations to be described
- discussing the relationships identified and realising that each one is adequate for describing any one of the analysed situations, that is for describing the motion of any object provided that it travels at constant speed

The operational definition of v:
- reinterpreting the relationship $v = d/t$ as the operational definition of speed in the case of uniform motion
- measuring speed (of Polistil cars) from the measurements of traveled distances and corresponding times
- formulating and checking predictions about the results of a race between two Polistil cars

Figure 1 – *The sequence developed in the classroom*

amples of races between two real bodies moving at different speeds, and to give for each of them the corresponding distance travelled and time spent. The examples were constructed by the pupils themselves so that the motion of each of the two bodies could be easily compared, the distance or the time being kept the same for each pair (see table 1). In this way, the comparison between speeds was reduced to the comparison of the values either of dis-

A challenge for lifelong science understanding

tance or of time, the same strategy spontaneously used by the pupils when describing their runs.

SAME TIME		SAME DISTANCE	
Boy A 1 minute 500 m	Boy B 1 minute 50 m	Luigi 100 m 20 s	Roberto 100 m 40 s
Red car 2 minutes 10 km	White car 2 minutes 5 km	Green car 5 km 1 minute	Blue car 5 km 5 minutes
Luca 1 minute 30 m	Leonardo 1 minute 60 m	Gianni 30 km 1 hour	Franco 30 km 2 hours
Ciao Motorcar 1 hour 30 km	Vespa Motorcar 1 hour 80 km	Michele 300 km 5 hours	Marco 300 km 10 hours

Table 1 – *The races analysed by the pupils (the examples were suggested on the spot by the pupils during the classroom work)*

The same examples, however, gave the adult the opportunity to stimulate the pupils to make a first step into disciplinary knowledge construction, by challenging them to solve the problem of how to compare the motion of bodies that had travelled different distances in different times. In figure 2 we report an excerpt of the discussion showing both the children's way of reasoning about the problem and the adult's ways of intervention.

As might be noticed, the problem posed by the adult (1), although non spontaneous for the pupils, is still solvable with the same strategy they used before, provided that a proportion is implicitly made to allow either times or distances to become equal for the two moving bodies (2). Children never studied proportions, but the reasoning required is still rather spontaneous involving operations that can be interpreted in their elementary meaning: multiplication as reiterated addition, division as reiterated subtraction or comparison between quantities of the same nature (from 4 to 20).

The interplay between the "facts" and their formal representations led one pupil to make links explicitly for the first time between the speed and "a couple" of values of distance and time and, at the same time, to identify the physical constraint at the basis of the correct way of reasoning: the necessity of assuming the uniformity of motion (20, 24).

The role of the adult was that of systematically 'throwing the ball back to the pupils', either by asking someone to make his thought as clear as possible (23), or emphasising the cognitive features of the issue posed by the child (4), or making the cognitive strategy used by him more explicit (21), or

(1) *Adult*: Well, we saw that the red car won the competition by travelling 10 km in 2 minutes, OK? Then we had a Vespa (motorcar) that won another race by travelling 80 km in one hour time. So I asked: in your opinion, had the competition been between the red car and the Vespa, which one would have gone faster?.. Now Matteo will say how .. which one he thinks to be faster, thoroughly explaining himself.

(2) *Mat*: Because, given that the Vespa in one hour did 80 km, whereas the red car in two minutes did 10, it was necessary to see how many they did in the same time.

(3) *Adult*: Wait a minute... is there anybody who has problems about what he said?

(...)

(4) *Adult*: The distances were different, 10 km and 80 km, but unfortunately so were the times, 2 minutes and 1 hour, so, if I understand well what Matteo did, Matteo said: given that the times were different, I tried to imagine what distance the red car would travel if it had 1 hour instead of only 2 minutes. Is it so? ... because, he says, if I know how long it goes for one hour it's just as before: same time, let's look how far each went.

(5) *Mat*: 600 km.

(6) *Adult*: so... one hour, how much is one hour?

(7) *Mat*: 60 minutes.

(...)

(8) *Mat*: To find the right multiplication I made 2+2+2 thirty times, that gives 2, 4, 6, 8 and it came to 60, repeating thirty times.

(9) *Adult*: So what you are saying is that 60 minutes is ...

(10) *Mat*: 30 times 2. (...) Then the result, that is 60, I multiplied it by 10.

(11) *Adult*: and what did you find?

(12) *Mat*: 600.

(13) *Adult*: I doubt it...

(14) *Fra*: me too.

(15) *Others*: me too!

(16) *Fra*: Because he did 30 x 2 and then the result x 10, that is 60 x 10 that is 600. But this wasn't what he had to do.

(17) *Adult*: What was it then? And why?

(18) *Fra*: Well, 60 divided by 2 equals 30 , times 10 gives 300.

(19) *Adult*: What does this 60 divided by 2 mean?

(20) *Fra*: If the red car continued to go with the same "rule" until it got to one hour, for example, each two minutes it went 10 km, it would get... it would do 60 divided by 2, if it continued with the same rule... it should do 60 divided by 2.

(21) *Adult*: That is when I divide 60 by 2 I am finding how much bigger one hour is compared with 2 minutes. So you say: if it continues to travel 10 km, every 2 minutes 10 km, since in one hour there are 30 times 2 minutes, then you say the distance travelled by the red car should be 30 times longer, in one hour it should be 30 times longer than in 2 minutes. Do you agree Matteo?

(22) *Mat*: yes.

(23) *Adult*: Please Francesca, say what you mean by 'given that the same rule is kept'.

(24) *Fra*: I mean, it means that... given that each two minutes it continues to travel the same road.

Figure 2. *Comparing the motion of bodies that travel different distances in different times*

A challenge for lifelong science understanding

checking that the other pupils were also following his line of thought (3), or simply by expressing a doubt and letting some other pupil pick it up (13).

Indeed, by using the new ideas in working on the examples of table 1, the pupils were guided to reach a first systematisation of the results of their spontaneous strategy of reasoning. By doing this, they found out that in uniform motion the ratios between two distances and between the corresponding times are equal, and that when comparing uniform motions with different speed, the ratios both between the distances travelled in the same time and between the times spent on travelling the same distance, are constant.

On this basis, the idea that a relationship exists between the variables v, d and t was developed. For this purpose, the problem was posed by the adult in terms of how individual results could be written in a more general form that could account for all situations. This is not the kind of problem the pupils would spontaneously see: it is a typical problem that physicists set to themselves, and that characterises physics as a discipline.

The pupils, starting from concrete examples, struggled to give a meaning

---

(1) *Ali*: So we had 120 km as the distance travelled by the red car and the speed at 300 km per hour... and we made ... we had to find the time.
(...)
(2) *Ve*: and we had to find the time.
(3) *Ali*: We had to find 24 minutes (which was the time written in the table). So the first time we made 300 times 120, that is a multiplication.
(4) *Ve*: The distance times the speed.
(5) *Ali*: The distance times the speed and we thought the time would come out, but we got 36.000 and this was wrong and we checked that it was to high.
(6) *Adult*: Indeed it is not true that speed times distance gives the time, what would it mean?... It would mean that, being equal the distance travelled, the higher the speed the longer the time and this is not true.
(7) *Ve*: Since the first was wrong we tried another. We then tried the speed, that is 300 km per hour...
(8) *Ali*: divided by the distance.
(9) *Ve*: divided by the distance that is 120 km. And we got 25. (...) Then we said: we may have got the calculation wrong and we tried again but it always came 25. So we tried another.
(...)
(10) *Ali*: Third case: again with the distance of 120 km and the speed of 300 km per hour, we made distance divided by speed that is 120/300 and got 0.4 hours and then we changed this into minutes and we got 24.
(11) *Adult*: So distance divided by speed, you said OK it works, distance divided by speed gives the time.

Figure 3. *Finding and checking relationships among the variables v, d and t.*

to the formalisation of the concept of speed by identifying how the three variables can be connected by "a formula". It is by checking that facts are in agreement with the formal link between variables that the pupils could give a meaning to the relation itself (see figure 3).

This appears to be a crucial step for giving a meaning to the final problem posed by the adult of how many forms the relationship among the three variables can take.

Indeed, when one pupil put forward the relationship $d/t=v$, that is the operational definition of speed, another pupil (Eli, see figure 4) used rather general arguments in order to justify its correctness. Her arguments reveal her awareness of the general properties that the relationship should satisfy: to be valid in all situations of uniform motion (1), to be in agreement with the conditions of uniform motion (3), to be defined unambiguously (11).

At the end of the learning path all pupils except two were able to use the relationship between d, t and v, though mastering it at different levels:

- 3 pupils were able to solve problems in simple situations only;
- 5 pupils were able to solve problems concerning complex situations also;
- 7 pupils were also able to invent problematic situations and solve them.

---

(1) *Eli*: I had found this too. I shall take the simplest example, given that it has to work for all, with all numbers.

(2) *Adult*: OK, this is an important point.

(3) *Eli*: If the distance is 2 km and the time 2 hours, if I ... that is I have proceeded by exclusion, because if I make distance times time it cannot... it cannot be right, because distance 2 km, time 2 hours, the speed would be 4 km per hour, but the speed should come out... cannot come out higher, it has to be always the same, I say...

(4) *Adult*: Elisa is finding another wrong formula and throws it away by argument. Have you understood her argument?... She says: if, in order to find the speed, I take the distance...

(5) *Eli*: and multiply it with the time.

(6) *Adult*: What would I find?

(7) *Eli*: I would find the speed, so to say, altered, because...

(8) *Adult*: she would find a speed that grows as time passes...
(...)

(9) *Eli*: given that multiplying doesn't work, it must be...

(10) *Adult*: Ha, you say: first I try by categories of operations... but there are two divisions...

(11) *Eli*: Indeed I tried the first that came to me, of those two, then as I found one and saw that it worked, I thought... since these are two different ways, if this works, the other cannot work too, so I haven't even tried.

Figure 4. *Identifying the general features of the relationships*

A challenge for lifelong science understanding

Children's perception of the classroom work

The children's perception of the classroom work was monitored using two individual interviews with all of them (one after the first part of the classroom work and one at the end), and a written questionnaire at the end of the classroom work.

The first interview was based on the discussion of the solutions to problems given to the pupils as homework, the second one aimed at investigating whether they appreciated various features of the new description of motion.

The questionnaire aimed at collecting the pupils' evaluations of the teaching pathway, and of the teaching and learning processes developed in the classroom.

Being concerned with the balance of the teaching experiment, we shall base our considerations on the pupils' answers to the second interview and to the questionnaire, whose features are reported in appendices 1 and 2.

All pupils except two, who had been pointed out by the teacher as finding learning difficult, were able to express their opinions on the two issues. We have selected some comments to illustrate the main issues raised by the group of pupils.

During the interview almost all pupils recognised that they had developed new ideas about motion, often admitting that they had never cared too much about motion before.

> *Ve*: Well, for example, that time we used the red car, the white car, the motorcar (referring to table 1)... that is we found the distance, the speed, the time and... they were... they were, for me, ideas I did not have before, of finding the speed and the time, because I really didn't care.
>
> *Ste*: ... I mean I did not think about time, no I said... I mean I did not put either speed or distance, I mean I really didn't think about that. I did, for example, I wrote a text this way, I said, for example, I went to see my Aunt, but it isn't that I wrote, so to say, distance, how long it takes, the speed. That's why it is a new way for me.

Some also accounted for the idea they had of physics and how this has changed.

Ma: That to find the time, for example, one has to divide the distance by the speed, I had never thought before. (...) because I thought that motion...wasn't a subject, I did not think that physics dealt with motion, I thought it was something more... how do you say? I did not think it dealt with motion.

Some pupils noticed that physics is able to give a quantitative description of motion in comparison with the qualitative nature of their common-sense descriptions, also emphasising the practical advantages of this new knowledge:

El: I mean that before, when (I looked at) the motion of my school mates who ran, I only thought that one was going faster, very fast. Then those three words: distance, speed and time, I understood better that every speed... I mean every distance, has its own time and speed.

Di: When I see someone passing, I say that he is going fast, I only say that because I don't even know the speed, whereas here we can describe how fast he goes...

Gi: Because I couldn't before... know the speed just by knowing the distance and the time, because I did not know the formulae to find them, so when I learnt them, then with only two data one can get the third too.

Eli in particular gave a rather accurate account of the meaning of the new description of motion as a new, powerful language for knowing and for communicating:

Eli: Because now, if I want to make someone understand... I mean, if I really want to describe something about motion, those words, I mean, I can use them, they are part of my vocabulary, whereas before... (I knew) nothing of these new ideas.

It is important to note that although the same ideas were expressed by different pupils, each pupil appeared to have been struck by some particular aspect of the work, that appears to resonate with his/her personal view of school, of study, of the discipline, and of his/her interactions with peers and with the teacher. In other words the picture that emerges by looking at all the answers of the different pupils does not portray the opinions of any one

A challenge for lifelong science understanding

of them but rather gives an insight into the range of possible meanings stimulated by the classroom work.

The possibility to find the work interesting and enjoyable in different ways was strictly linked to the flexibility and richness of the learning environment that proved to be able to respond to individual cognitive needs and to make the best of individual cognitive abilities. This is documented by the comments written by the pupils in their answers to the questionnaire and by the variety of different justifications they gave. The effort required by pupils was acknowledged by someone and this was positively judged as stimulating interest, despite the difficulties that must be tackled, because the pupils felt themselves to be adequately supported.

> Eli: The most interesting activity was when we found the three formulae that were linked with one another, because I really did not know them; but they have also been the most difficult exactly because I did not know them and I struggled to understand. This way of studying is easier if compared with reading in a book, because I learnt much more from the voice, the sheets and the real examples than in a book. For me it's well done! Because we thought together, we solved the problems that we met, we made examples to understand better and finally we revised the result that had come out.

Many pupils emphasised the affective component of the style of work, by using terms like '*simpatico*', '*nice*', '*amusing*'. *Ve* in particular linked this to the active role played by the pupils (*'it's a cheerful way of discovering new things for those who do not know them'*), *Mag* to the higher level of understanding (*'something studied may not become engraved in one's memory, whereas something 'studied' in an amusing way leaves a trace in one's memory'*), *Ma* to working in groups (*'not only did you make us solve exercises about motion, but you also taught us to work in groups'*), *Sa* to the importance of seeing how things work (*'reading in a book I may know but I cannot see how to make the motion'*).

Others emphasised the role played by the adult in supporting understanding, *Eli* from a strictly personal point of view, *Gi* also from the point of view of the teacher.

> Eli: Studying this way is simpler because if a child does not understand there is always someone there to explain to him. It's very amusing,

though sometimes to repeat what has already been done may become for many children, including myself, a little boring.

Gi: This way is correct because at the end of the work the teacher knows whether the pupil understood or not and, according to the result, the teacher knows whether to repeat the lesson in order to make him better understand.

The whole picture of the pupils' comments indicates, in our opinion, that the main aspects both of the nature of disciplinary knowledge and of the learning environment created in the classroom have been appreciated by the pupils, though not every aspect by all the pupils nor at the same level of awareness for all of them. The teaching experiment shows then that primary pupils are sensitive also to metalearning analysis and can be stimulated to reflect on the processes and products of knowledge construction. Even a pupil who had had difficulties in getting involved with the classroom activities appeared to have realised that the work done had to do with something more than simply learning by heart about motion.

Fe: I have learnt to respect studying and to reread when one hasn't understood well.

The "laboratory of facts and ideas": a challenge for the teacher

A salient feature of the work in this kind of laboratory is that the path to follow cannot be designed in detail *a priori*: individual content knowledge construction is not the repetition of a fixed path that somebody else, however experienced, decided to be most efficient. The adult who leads the classroom work knows where to go, but the steps on the way need to be identified each time by a process of continuous comparison between the pupils' ways of looking, and the concepts and methods peculiar to the disciplinary perspective in order to fit the former with the latter. The resulting path is not a linear one, it goes backward and forward several times according to the interplay of facts and ideas which is at the basis of laboratory work: the shape of it is peculiar to the features of the class and to the pupils' interactions with adults and their peers.

The "facts" in this kind of laboratory can be of a different nature. The first approach to the investigation is always based on a phenomenological study of non schematised situations, stimulated by problems to be solved. The facts are then modified in order to identify simpler contexts to analyse, drawing first level conclusions and then going back to the more complex situations to be re-analysed by means of the newly constructed conceptual tools.

The collective construction of new cognitive tools is indeed the main driving force of the labwork: each pupil should have the chance to contribute to this but each one will benefit from it in different forms and at different levels. It is important not to assume that the individual paths and the collective one can be simply identified. Therefore tasks to be performed individually are a necessary part of the activities carried out in the laboratory. They allow the adult to follow the individual development of pupils, and also allow each pupil to keep track of his/her own path within the flow of collective work.

To promote collective construction of knowledge is not an easy task. In the experiment described in this paper, great effort was put into developing child to child dialogue and in reducing adult centred interaction.

The first excerpt (figure 2) shows the importance of the role of the adult in inviting the pupils to make their ideas and proposals explicit and in guiding the pupils to reformulate them in a format nearer to the accepted disciplinary one, though still meaningful to the pupils. Another example of this important role is when the adult summarises the path followed by the pupils in the construction of new ideas, emphasising the development of the main ideas and the necessary steps in the process. The reconstruction of the path is indeed important for the pupils to reflect on their knowledge and on the process of construction itself and to become aware of the nature of the knowledge constructed in comparison with their previous knowledge.

"The challenge for the science teacher involves helping the students to make sense of scientific ways of knowing in terms of their existing knowledge, and then to differentiate these two ways of knowing" (Leach & Scott, 1995, p.49). The "laboratory" is the context for making sense of new ways of looking *while* realising in what way they are different from the spontaneous ones.

The "laboratory of facts and ideas": an investment for the future

The teaching experiment has shown that the "laboratory of facts and ideas" is a special learning environment that can be put into practice and can work efficiently at primary school level.

It also appeared that at the age of the pupils in the study (9-10 years) some of the children have already been left out of the game: an earlier involvement in meaningful science education would perhaps give them too the cultural opportunities that their family and social context cannot provide.

Indeed labwork can start well before the 5th grade. "Children come to science, at age five, already able to observe, classify, hypothesise, predict, compare 'fairly', and so on, with high levels of skill in contexts where they see the purpose in doing so" (Millar, 1996, p.15). The "laboratory" can provide a meaningful context where the children's skills are applied to problems peculiar to science and acknowledged to be so (see for example Gagliardi, 1987).

The teaching experiment has also shown that the time required to deal with a particular piece of knowledge is much longer than in a more traditional setting. A large percentage of this time was spent in order to allow the pupils to adjust their mind to the kind of "game" they were invited to play. The fact that almost all pupils managed to resonate with the teaching proposal shows, in our opinion, that this time must be seen as an investment for the future: we believe future activities will profit from the knowledge and abilities constructed (ways of looking, of doing, of communicating, etc.), and that the development of a positive attitude toward learning science (the desire and the pleasure of learning, the awareness of one's own ability to understand through investigation) will be a solid basis for learning and enjoying physics at higher levels of education.

If one believes that science should be a relevant component of the culture of every citizen, then it is in the earlier years of schooling that the foundations for this must be laid.

References

Arcà M., Guidoni P. & Mazzoli P. (1984) – Structures of understanding at the root of science education. Part II: Meanings for formalization – *European Journal of Science Education* 6 (4), 311-319.

Gagliardi M. (1987) – La modélisation et les differents codes de representation symbolique à l'école élémentaire *[Modelling and different codes of symbolic representation in elementary school]*- in A. Giordan & J. L. Martinand (Editeurs) 'Modèles et simulation – Actes des IX^{emes} Journées Internationales sur l'éducation scientifique' – Centre Jean Franco, Chamonix.

Grimellini Tomasini N., Gandolfi E. & Pecori Balandi B. (1990) – Teaching strategies and conceptual change: Sinking and floating at elementary school level – Paper presented at the AERA Annual Meeting, Boston, *Eric/Resources in Education*, ED 326 428.

Grimellini Tomasini N., Pecori Balandi B. & Gagliardi M. (1992) – Reasoning, development and deep restructuring – Paper presented at the AERA Annual Meeting, S. Francisco, *Eric/Resources in Education*, ED 347 183.

Leach J. & Scott P. (1995) – The demands of learning science concepts. Issues of theory and practice – *School Science Review* 76 (277), 47-51.

Millar R. (1996) – Towards a science curriculum for public understanding – *School Science Review* 77 (280), 7-18.

Piaget J. (1946) – *Les notions de mouvement et de vitesse chez l'enfant [Children's notions about movement and speed]*- Ed. PUF, Paris.

Scott P. (1996) – Social Interactions and Personal Meaning Making in Secondary Science Classrooms – in G. Welfort, J. Osborne & P. Scott (Eds.) *'Research in Science Education in Europe'*, Falmer Press, London.

M. Gagliardi, N. Grimellini Tomasini, B. Pecori

Appendix 1: Interview grid

(The interviewer always checked that the questions posed were clear to the pupil, and in cases where they were not she helped the pupil to understand the questions by reformulating them.)

1) About students' way of looking at motion

Question:
The aim of the classroom work was to observe and describe the motion of bodies.
Do you think that the work has modified your way of looking at motion from the one you had at the beginning?
In the case of a positive answer:
How do you think your way of looking at motion has changed?
In the case of a negative answer:
Please tell me in what sense it has not changed.

2) About language

Question 1:
Did you learn some new words during the classroom work about motion? Words that you did not know before?
In the case of a positive answer:
Which ones?

Question 2:
Did you find new meanings for words that you knew already? Did you use any words that you knew already in a different way?
In the case of a positive answer:
Which words? In which different way did you used them?

Question 3:
Of all these words which ones you think are more important in the description of motion? Why?
Further question (if appropriate):

A challenge for lifelong science understanding

In your opinion do these new words correspond to new ideas? What are these new ideas?

Question 4:
During the classroom work we saw that a link exists between three quantities: distance, speed and time. The link was expressed in three different forms:

$d = v.t \qquad v = d/t \qquad t = d/v$

Do you think that these relations can be looked at as a new way of describing motion?
In the case of a positive answer:
How do you think they can?
In the case of a negative answer:
Why do you think they cannot?

M. Gagliardi, N. Grimellini Tomasini, B. Pecori

Appendix 2: Questionnaire about the study of motion

The pupils were given a brief summary of the classroom activities that they had followed, and a schematic description of the way in which classroom work had developed through interaction with facts and ideas. They were required to give written answers to the following questions:

1. Is the description of the classroom activities clear and does it correspond, in your opinion, to the work we did together? Would you add or remove anything? If yes, what?
2. Which activities were the most interesting to you; which ones were more difficult?
3. Do you think that this way of studying motion makes it easier or more difficult to understand the topic, compared to studying it using the textbook? Why?
4. What is your opinion about the way we worked together?

A longitudinal study on labwork and 10-12-year-olds' development of the concepts of transformation of matter

Olle Eskilsson

The same group of pupils are being followed for 2-3 years, in order to study how labwork influences their understanding of everyday phenomena involving transformations of matter. There are forty pupils in the group, born in 1987.

This paper describes the study to date. The classes have very little experience of practical work in science. A particular focus was upon the first lesson, where I investigated how the children use concrete material, how they link this material to experiences of everyday phenomena, and how practical work has challenged and developed their conceptions. When grouping different substances, the pupils arranged the substances in groups and moved the substances between the groups as they discussed the relations between the substances. Their everyday experiences, rather than scientific concepts, guided them in grouping the substances.

Background

In Swedish primary schools, teaching tends to focus upon different phenomena rather than why they occur. In secondary schools, many pupils therefore do not make the great step to using theoretical models and concepts far from their reality. Until recently, it was common for the teacher to guide practical work very strictly. The focus of this study was upon how everyday phenomena can stimulate the curiosity of the pupils, especially during their labwork.

Olle Eskilsson

In school it is not possible use strict scientific concepts. Rather, it is necessary to choose concepts of a lower level than the scientific. Often we can use concepts that correspond to the pupils' everyday language. Notions about concept levels have been used in two ways. Firstly, the conceptual structure has been analysed, and secondly, the cognitive demands on the learner has been considered. Black and Harlen (1993) discuss a spectrum of concept levels. The children's everyday concepts are very limited. On the other hand, we have the scientific concept level which includes a small number of concepts, each applicable to a wide range of situations. Teachers need to know both these levels in order to help pupils in refining their concepts. Koulaidis (1995) describes three different knowledge frames of Science/Chemistry Education: Scientific Knowledge, School Science and Children's Ideas. An important step forward would be to try to describe these concept levels and their differences, and to identify ways of letting pupils make transitions between these levels (Lijnse, 1990). On the scientific level, we often have the big 'supertheory' as an explanatory umbrella, but the everyday 'theories' of young people are fragmentary and local (Claxton, 1993).

Studies about how younger children describe different types of transformations of matter in everyday phenomena were reviewed. Researchers in science education who have studied the use of a low-level particulate model have come to different conclusions. Some of them argue that young children are ready to use this model when they talk about everyday phenomena (Eskilsson, 1997; Novak & Musonda, 1991; Nussbaum, 1993). Millar argues that the particulate theory could be introduced in connection with solids but that it is more demanding to use the model in the case of gases (Millar, 1990). When using a low-level particle model, children often attribute macroscopic properties to the particles (Lee et al. 1993).

Richard White (in press) argues that *"a longitudinal application of Piaget's cohort studies of young children would be exceptionally valuable"*. Only 34 examples of longitudinal studies in science education can be found in the period 1963 – 1987, and since that time there has been little change. Some of the longitudinal studies where researchers have met the same pupils over several years and discussed concepts in the same context were reviewed. Helldén (1995) followed the same pupils' understanding of the organic conditions required for life and decomposition amongst pupils aged 9 to age 15. Novak and Musonda (1991) used two units of audio tutorial science lessons, one in first and one in second grade. The pupils were then interviewed periodically to assess changes in their science understanding from grade one to grade twelve. One of the two basic concepts in their study was the idea that matter has a particulate nature (Novak & Musonda, 1991). In longitudinal studies, the conceptual development of individuals, as well as the influence of different experiences on pupils' thinking, can be followed.

Labwork has been a central part of school science for many years. Pupils' learning in labwork and pupils' concepts of the nature of science have been described. Lemke (1993) has studied the role of language in science education. Lemke argues that science in the laboratory is just as different from science in the classroom as it is from other subjects. Lemke also argues that there are not many parallels to science laboratory work in other core academic subjects:

> In the laboratory, students talk science to each other to guide themselves through prescribed experimental procedures, to decide what to do when something seems to have gone wrong, and to write up notes on what they have done. (Lemke, 1993; p 157)

Many science teachers regard hands-on experiments as the heart of science learning. Hodson (1993) recommends a more critical approach to practical work in school science. He questions whether pupils will be motivated by practical work, whether it helps them to understand scientific concepts better, and whether "scientific attitudes" will be fostered by practical work. Hofstein (1988) gives an historical summary of the aims for, and problems with, practical work in school science. Watson (1995) has studied the effect of practical work on students' understanding of combustion. Pupils tend to give descriptions of phenomena rather than explanations for them. Most children do not use their experiences from practical work to modify their concepts of combustion. Watson argues that pupils develop a better understanding if practical work concludes with a discussion of different models of explanations. Sjøberg (1990) discusses the aim of labwork in school science and one of his conclusions is that it is uncertain whether practical work helps pupils to learn scientific concepts and theories.

Olle Eskilsson

Research question

The preliminary research question for the study is:
How does practical work influence the development of pupils' understanding of transformation of matter?
This will be addressed by considering the following questions:

- Which models of matter do 10-12-year-olds have and how do these models develop?
- Which concepts of matter do the pupils use when they explain everyday phenomena?
- How does practical work effect the development of their concepts?

Teaching and learning perspective used in the study

Ausubel (1968) emphasised the great importance of considering what pupils already know in planning teaching. In this study, data are collected around everyday phenomena that the pupils recognise. The purpose of discussions, interviews and practical work was to challenge pupils' conceptions in three steps:

- Interviews with individual students;
- Practical work in small groups to reach consensus;
- Whole class discussions introducing new concepts.

In order to complete practical work in science, it is necessary to talk science (Lemke, 1993). In both small groups and whole class discussions, pupils were therefore required to compare different explanations of phenomena.
It can be difficult for children to realise the difference between everyday and scientific concepts when they encounter everyday situations in interviews. The discussion of everyday phenomena triggers students to respond with everyday knowledge.

A longitudinal study

Design

The same group of pupils will be followed for 3 years, 1997-2000. Each child was given a pre-test so that the development of each child can be followed. Every term, the pupils will follow an instructional unit of three lessons, and each pupil will be interviewed about two months after instruction. During the last term of the project, the pupils will do a post-test with the same questions as the pre-test. During each instructional unit, data will be collected to see whether the practical work stimulates the children to ask questions about the phenomena they study. A particle model for matter will be introduced, and used in discussing various phenomena. Fig.1 summarises the design of the study. In this project, I acted as both teacher and researcher.

The sample includes pupils in five classes, drawn from two different schools. The pupils were born between1986 and 1988. The sample for the interview study includes about 40 pupils born in 1987.

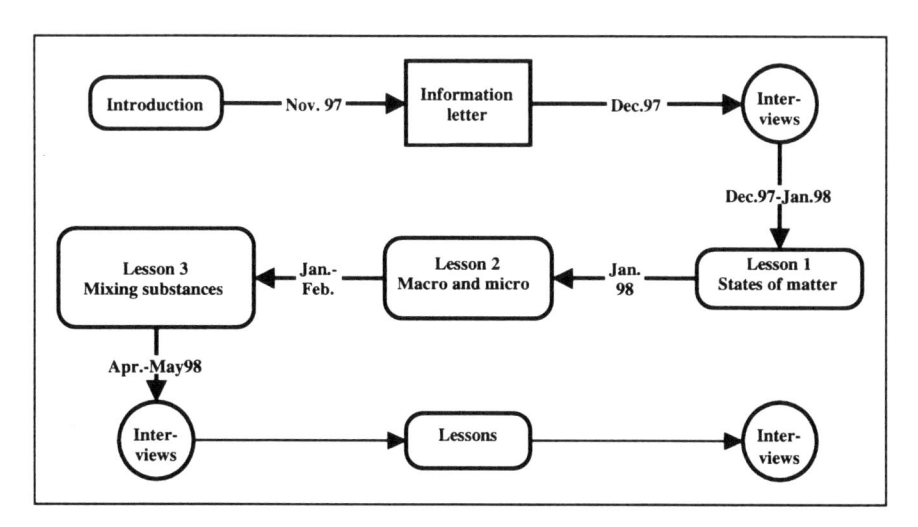

Figure 1. *The longitudinal study*

The revised clinical interview method was used (Pines et al. 1978). Experience from an earlier study suggests that this method provides good information on pupils' conceptions (Eskilsson, 1997) because questions can be rephrased and answers followed up with new questions to avoid misunderstandings. The pupils are presented with 3 or 4 situations familiar from everyday life. The situations are illustrated with different objects such as a burning candle or a glass jar with soil. The focus of analysis is upon which concepts the pupils use and how they use them. All interviews are to be audiotaped and transcribed. Analysis will focus upon trends in the group as a whole, as well as trends in individual students' thinking.

The study will focus upon concepts such as states of matter, the water cycle and chemical reactions. During the instructional units, the pupils will work in groups of four. Each group will be given the same problem to solve, and afterwards their results will be discussed. Everyday objects such as spoons, plastic mugs, magnifying glasses, cardboard etc. will be used. The problems involve classifying different materials, melting ice, studying macroscopic particles in sugar and wood, and some chemical reactions that result in new colours. A simplified particle model for matter will be introduced, and this will be used by the teacher in discussions. Students will also be encouraged to use this in discussions. All teaching will be videotaped. Videorecording will focus on group discussion, and these discussions will be analysed for both content, the questions the pupils ask each other, and the decisions reached in the group in order to see how the practical work challenges and develops pupils' conceptions. This record of group discussion may also be used in future interviews.

Some teaching sequences in the study

Fig 2. shows some of the concepts focused upon on in the study, and summarises progress to date. All the pupils described how rain is formed, in a paper-and-pencil task. The purpose of analysis was to characterise how the pupils described the phase changes in water during the cycle. In the first interview, students were shown a sealed glass jar with soil on the bottom. The jar had small drops of water on the internal walls and students were asked where the water had come from. One task in the first lesson was to melt an ice cube as quickly as possible. Students were also told that water in the water tap, snow on the ground, and water in the air are all water, but in different states. In the second lesson students were introduced to ideas about macroscopic and microscopic particles in solids, liquids and gases, and in the second interview discussion focused upon what happens to water that is left on the kitchen sink.

A longitudinal study

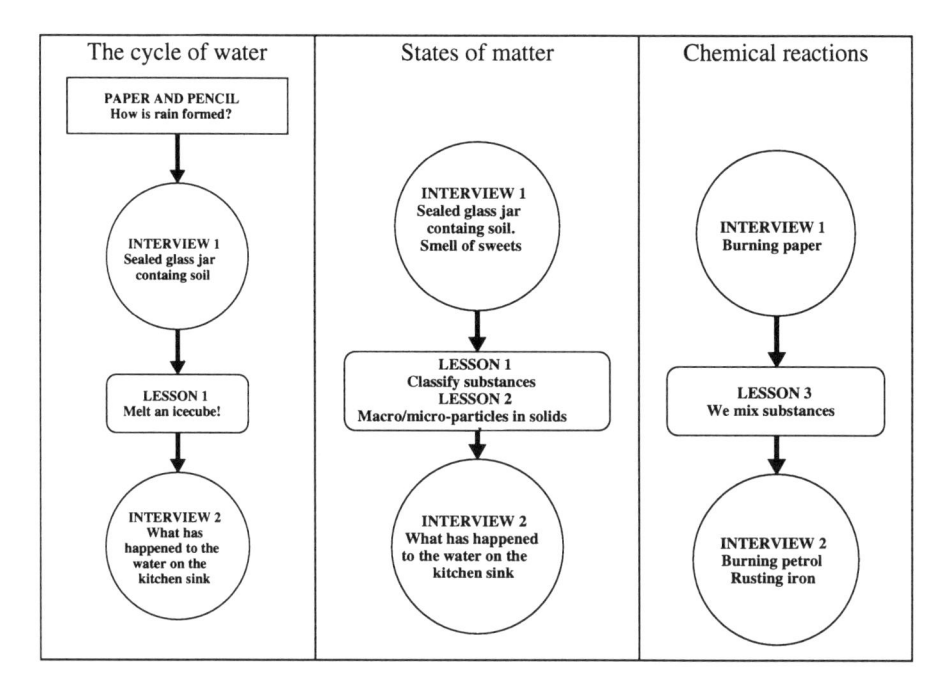

The cycle of water	States of matter	Chemical reactions
PAPER AND PENCIL How is rain formed?		
INTERVIEW 1 Sealed glass jar containing soil	**INTERVIEW 1** Sealed glass jar containing soil. Smell of sweets	**INTERVIEW 1** Burning paper
LESSON 1 Melt an icecube!	**LESSON 1** Classify substances **LESSON 2** Macro/micro-particles in solids	**LESSON 3** We mix substances
INTERVIEW 2 What has happened to the water on the kitchen sink	**INTERVIEW 2** What has happened to the water on the kitchen sink	**INTERVIEW 2** Burning petrol Rusting iron

Figure 2. *Some concepts in the study*

Interview contexts were chosen in order to investigate how the children describe chemical reactions that they meet in everyday situations. Combustion was one example of a chemical reaction used in the interviews. In the first interview, pupils talked about what they thought would happen if a piece of paper was set on fire, the role of oxygen, and if anything would remain. They were also asked about what would happen if we put a jar over a burning candle, and where the water on the jar comes from. In the last lesson, various substances were mixed to see what would happen (e.g. sugar and red cayenne, lingonberry and basic washing detergent, lead nitrate and potassium iodide). These substances were chosen because it was easy to see from the colour if new substances had been formed or not. In the second interview, pupils were asked what would happen when we set fire to some drops of petrol, what would happen when we put a beaker over the burning petrol, what interactions occur between petrol and oxygen, and where the water comes from.

As already stated, one aim of analysis was to see if a presentation of a simple particle model can help young children to understand these everyday phenomena in a more scientific way. In the first interview, pupils were asked to explain how the water came from the soil to the walls of the jar. Then they were asked what it might look like if we had 'magic glasses' so that they could see everything in the jar (following Nussbaum, 1985). From their responses, it was clear that some pupils had heard of atoms and molecules prior to teaching. In the second lesson, pupils first studied smaller and smaller macroscopic particles of sugar with a magnifying glass. Then, the sugar was put into water and pupils were asked where the sugar was after mixing with the water. Drama was used to describe this situation, and at the same time, a theory of small particles that build up sugar was introduced. Particles were referred to as 'molecules'. Molecules were talked about in water, in their hair and so on. During the drama, we showed what happens in terms of particles when sugar is dissolved in water, and when ice melts and water evaporates. During the second interview, analysis focused on how this model was used. If pupils didn't use it, they were asked if they remembered what molecules are.

In this paper, I will show some examples from the work in the first series of lessons. My idea was to give the groups of 4-5 pupils problems to discuss, followed by a class discussion where the groups explained the conclusions that they had drawn. After comparing the different solutions proposed by the pupils, I summarised and presented some models we use in science, on a level that seemed appropriate given their initial conceptions.

The first problem for the pupils was to group some substances I had put together in separate plastic cups. The substances included water, toothpaste, sour milk, potato flour, juice, a balloon with air, woollen cloth, wood, aluminium, a piece of chalk, a flower, a straw in a cup with water ('the bubbles'). All groups discussed and compared the substances in the cups and considered different ways of classifying them according to similarities and differences.

Preliminary results
What happens in the groups?

As only one video camera was available in each classroom, it was not possible to follow the whole discussion in every group. [In the transcripts, my comments are given in brackets.]

Karl: The flower and water fit together.
Cecilia: The bubbles in water fit together. It's carbon dioxide.
Karl: Sugar and that (juice) fits together. (Juice is sweet)

Emma's group concentrated on their prior experiences of the substances. They were discussing their ideas and asking each other questions during their work.

Henrik: You can eat them. This one burns quite fast (aluminium).
Charlotte: It's melting yes. Can chalk melt? (Charlotte asks Henrik)
Emma: Yes, chalk can be melted.
Charlotte: Can we eat toothpaste? (Edible things had been discussed before).
Henrik: Can we make a group of those we use in real life?

One group used chains in their explanations. They looked for connections between one subject and another, and then a connection to the next, and so on.

Petri: These you can eat.
Sofie: Toothpaste?
Johannes: The straw is made of plastic like a toothbrush: (straw in water and toothpaste is placed in a group.)
Angelica: I will have toothpaste in the same group as water.
Petri: Yes sometimes you gargle your throat with water.
Angelica: I will have the potato meal in the water group.
The others: No.
(Johannes then takes the chalk and Angelica the aluminium)
Johannes: Aluminium is a metal and the table has table legs of metal. Wood and metal?
Angelica: Yes

Classification and pupils' explanations during teaching

After the group work, each group reported on what they had come up with and what they had found in common amongst the substances. In analysis, I looked for a structure in the pupils' ideas of similarities between substances, drawing upon an idea used by Krnel (1995) in a study of how pupils aged 9, 11 and 13 classify everyday materials. Overall, there were 24 groups in the classes. Table 1 shows on the one hand the number of times a given explanation is used, and on the other hand the number of groups that used the explanation.

Explanations involve	number of times used	groups
state of matter (liquid, solid, gas)	0	0
kinds of matter (air, water, wood)	26	22
action or purpose (keeping, for constructing, for eating, need water)	22	15
relation (natural, are made of...)	9	9
perception (look like, same colour,...)	12	10
sequences (water contains oxygen, water wets wood)	21	16

Table 1. *Motivations for classifying*

None of the groups referred to states of matter in their classifications. However, most of the groups placed the two examples of gases in the same group and called it 'air'. This way of classifying was categorised as 'kinds of matter. "Flowers and trees need water, and cloth can 'suck' water, these three can suck water" is categorised as 'action or purpose', "Flower, tree, plants, sour milk, and potato flour come from nature" as 'relation', "Toothpaste and sour milk are thick" as 'perception' and "Juice is drinkable like water, that contains oxygen and there is oxygen in the balloon" as 'sequence'. Most groups discussed the relationship of the substance to water. These discussions were often coded in 'action or purpose' and 'sequences'. This classification of substances is based on pupils' models of matter prior to teaching, pupils' everyday experiences, and discussions in their groups. It will be used as a reference when studying conceptual development.

Sequences

Many groups discussed relationships between the substances during the lesson. They often described this in terms of chains between, for example, juice and water and then water and flowers. Then they placed juice, water and flower in the same group. In Figure 3-6 some of these 'chains' of links between subjects in the same group are identified.

A longitudinal study

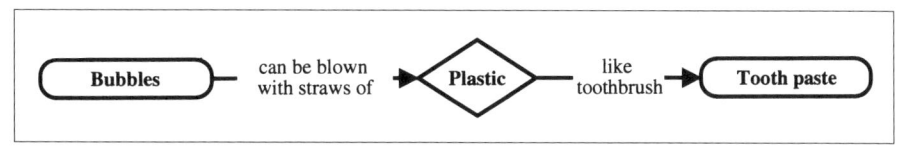

Figure 3. *Plastics in straws and in toothbrushes*

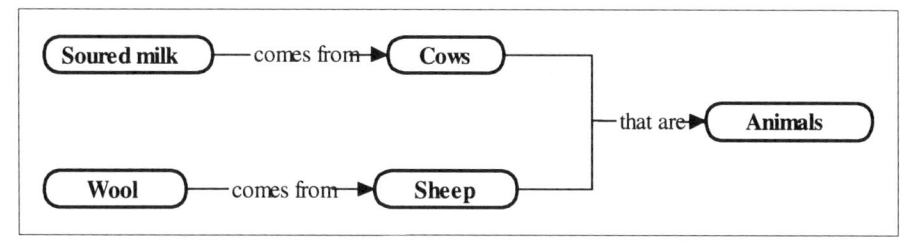

Figure 4 *Associations to animals*

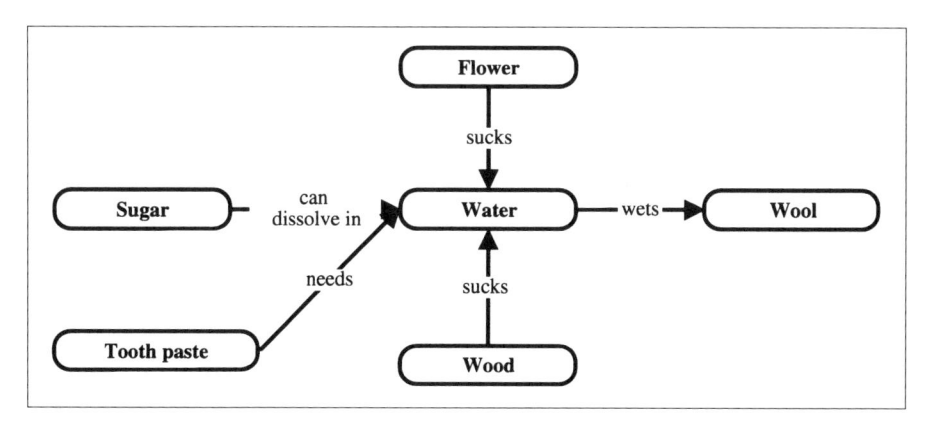

Figure 5 *Links to water*

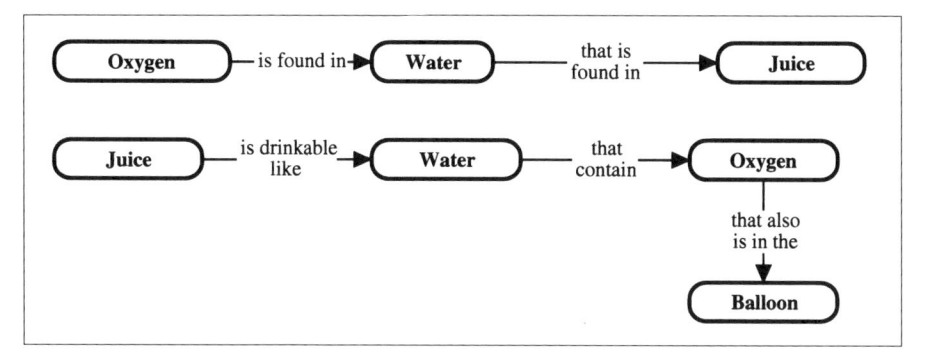

Figure 6 *Oxygen in water and in air*

239

Olle Eskilsson

Summing-up

After the groups had presented their ideas for classifying the substances, there was a class discussion of the different classifications, what they had in common, what was good and so on. The model of states of matter was presented, and used to classify the substances as solid, liquid, and gas. The Swedish word for liquid, *flytande*, can be used for powders too, and the distinction between powders and liquids was therefore made carefully. Pupils were presented with the idea that water can exist as a liquid, solid and gas, and this was linked to the question of 'rain' and the sealed jar with soil.

Who can melt an ice cube in the shortest time?

Every group was given an ice cube and a plastic bag. They were given the challenge to melt that cube as quickly as possible. They were not allowed to use water. The purpose of this activity was to engender another opportunity to discuss water in different states.

Some groups tried to warm the ice with their body heat and others with their 'warm clothes'. When one method didn't work they chose new methods. Some of the methods used in the five classes (a-e) during the competition are shown in Table 2.

Method	a	b	c	d	e	Sum
"Warm clothes"	×	×	×		×	4
Breath on the ice	×	×		×	×	4
Rub the ice cube with hands	×	×		×	×	4
Take the ice (in the bag) in the mouth	×	×	×		×	4
Warm it with your body	×		×	×	×	4
Smash the cube to pieces	×	×	×	×	×	5

Table 2. *Methods used to melt ice cube*

The groups then evaluated their methods and the different methods were compared. During this discussion, pupils were encouraged to talk about the phases of the water.

The practical work involving problem solving challenged pupils' conceptions and contributed to their conceptual development. Pupils' everyday experiences gave ideas for experiments that were then carried out.

Future priorities for the study

During the lessons, the pupils used materials in their attempts to solve the problems. The pupils could probably have solved the problems without using the materials, but using the materials stimulated the discussions in the groups. They used their everyday thinking when grouping the substances. Although some of them had the two examples of gases in a gas-group, none had a solid-liquid-gas classification. This corresponds with the findings of Krnel (1995). Many of the groups looked for links (to water, from animal) when they grouped the substances. The ideas used by pupils were fragmentary and local (Claxton, 1993).

After the whole class discussions, where the results from the different groups were presented, concepts on a higher level were introduced, e.g. after the groups' presentations of how they had grouped the substances, the concepts solid, liquid and gas were introduced. In order to find out how the pupils' models of matter develop over time and how practical work effects this development, students will complete more problem solving activities in groups, combined with practical work in future instructional units. Further individual interviews, group discussions, and whole class discussions will be used to see which concepts the pupils use when they explain everyday phenomena. The problems focused upon will be selected on the basis of outcomes from the interview studies and vice versa.

The pupils were not used to practical work, though they often work in groups. The practical problem solving challenged most of them and that led to lots of discussion. In future, I plan to analyse how the groups reach consensus, and how they use the concepts presented to them during practical work and interviews.

References

Ausubel, D. P., Novak, J. D., & Hanesian, H. (1978). *Education psychology: a cognitive view*. (2 nd. ed.). New York: Rinehart and Winston.

Black, P., & Harlen, W. (1993). How can we specify concepts for primary science? In P. J. Black & A. M. Lucas (Eds.), *Children's Informal Ideas in Science* (pp. 208-229). London: Routledge.

Claxton, G. (1993). Minitheories: a preliminary model for learning science. In P. J. Black & A. Lucas (Eds.), *Children's Informal Ideas in Science* (pp. 45-61). London: Routledge.

Olle Eskilsson

Eskilsson, O. (1997). *Materiabegreppet i läroplaner och klassrum* (Concepts of matter in curricula and classrooms): Kristianstad University.
Helldén, G. (1995). Environmental Education and Pupils' Conceptions of Matter. *Environmental Education Research, 1*(3), 267-277.
Hodson, D. (1993). Re-thinking Old Ways: Towards A More Critical Approach To Practical Work In School Science. *Studies in Science Education, 22*, 85-142.
Hofstein, A. (1988). Practical Work and Science Education II. In P. Fensham (Ed.), *Development and Dilemmas in Science Education*: The Falmer Press.
Koulaidis, V. (1995). Is Chemistry Education a Scientific Discipline? An Epistemological Perspective. In R. M. Janiuk (Ed.), *Research in Chemical Education and its Influence on Teaching Chemistry at School* (pp. 12-20). Lublin-Kazimierz: Federation of European Chemical Societies.
Krnel, D. (1995). Towards the Concept of Matter. In R. M. Juniak (Ed.), *Research in Chemical Education and its Influence on Teaching Chemistry at School* . Lublin-Kazimierz: Federation of European Chemical Societies.
Lee, O., Eichinger, D. C., Anderson, C. W., Berkheimer, G. D., & Blakeslee, T. D. (1993). Changing Middle School Students' Conceptions of Matter and Molecules. *Journal of Research in Science Education, 20*(3), 249-270.
Lemke, J. L. (1993). *Talking Science*: Ablex Publishing Corporation.
Lijnse, P. L. (1990). Macro-Micro: What to discuss? In P. L. Lijnse, P. Licht, W. de Vos, & A. J. Waarlo (Eds.), *Relating Macroscopic Phenomena to Microscopic Particles* (pp. pp 6-11). Utrecht: CD-ß Press.
Millar, R. (1990). Making Sense: What use are particle Ideas to Children. In P. L. Lijnse (Ed.), *Relating macroscopic phenomena to microscopic particles* (pp. 283-293). Utrecht: CD-ß Press.
Novak, J. D., & Musonda, D. (1991). A Twelve-Year Longitudinal Study of Science Concept Learning. *American Education Research Journal, 28*(No.1), 117-153.
Nussbaum, J. (1985). The particulate Nature of Matter in the Gaseous State. In R. Driver, E. Guense, & A. Thiberghien (Eds.), *Children's Ideas in Science* (pp. 124-144): Open University Press.
Nussbaum, J. (1993). Teaching about vacuum and particles, Why, When and How., *Third Misconception Seminar Proceedings* . Ithaca NY.
Pines, A. L., Novak, J. D., Posner, G. J., & VanKirk, J. (1978). *The Clinical Interview: A Method for Evaluating Cognitive Structure* (Curriculum, Series, Research Report No. 6). Ithaca, New York: Department of Education, College of Agriculture and Life Science.
Sjøberg, S. (1990). *Naturfagenes didaktikk* (Science Education). Oslo: Gyldendal, Norsk Forlag A/S.
Watson, R. (1995). The effect of practical work on students' understanding of combustion. *Journal of science teaching, 32*(5), 487-502.

White, R. T. (in press). The revolution in Research on Science Teaching. In V. Richardsson (Ed.), *Handbook in Research in Teaching*. New York: MacMillan.

Section 4

Practical work outside the laboratory

Introduction

In many countries, 'research like' activities, investigations or projects are carried out in school science education. These activities are often participatory and take place as fieldwork or laboratory work that is planned and managed by the teacher, but influenced by the students' interests and abilities. Although it is claimed that these activities are in some sense 'authentic', they are usually confined within the framework of the school or institution

The paper by Miki Dvir and David Chen, and the paper by Peter van Marion in this section report on research where activities are going beyond these confinements of schooling, although the approaches used are different.

Miki Dvir is currently participating in an Israeli project, where schools are provided with an 'educational greenhouse'. Hence the learning environment of the school is extended with an authentic research environment, where students at the secondary level are engaged in project work for developing scientific process skills. Although there seems not to be any measurable, significant change in student performances the activities are promising for future educational development. The educational greenhouse involves students in supervised authentic research activities expected to ensure enculturation into biological and agricultural research and goes beyond the borderline of what subjects within biological and agricultural studies can be taught.

The educational greenhouse represents a piece of the "real world" drawn into the learning environment of the school. In the project work reported by Peter van Marion students are going out into the "real world" by taking part in environmental investigations and research projects involving partners within the local environmental management, local politicians and researchers. The authenticity of these Norwegian projects is only in the area of scientific and environmental research, thought it also involves environmental management and political education. Consequently the activities have a wider scope then enculturation into scientific research. Also the problems, which arise are complex and beyond experiences in traditional schooling

Within the confinement of traditional schooling, activities are planned and carried out in a 'protected environment'. The arguments for the educational value of authentic activities, which go beyond the traditional culture of schooling, are very convincing. The problems which arise when dealing with 'the real world' are, however, also going beyond the problems experienced in tradition schooling. The authors of the two papers report these valuable experiences.

Albert Paulsen

The theoretical and practical aspects of actively involved inquiry learning in the Educational Greenhouse – a case study based analysis

Miki Dvir and David Chen

The objective of this paper is to present some aspects concerning the use of Educational Greenhouses (EG) in Israel. This inquiry related learning environment provides a platform for projects in biology and life sciences. The main issues presented are: a. The possible effect of students' involvement with authentic research activity on their understanding and interpretation of science process skills; and b. The realization of the EG's potential as a unique learning environment.

Preliminary observations relating to a variety of cognitive abilities, indicate that following this specific kind of exposure to active involvement in science, students' understanding and correct use of higher order scientific process skills, were unsatisfactory. In terms of the second issue, EGs are yet far from being fully integrated within secondary schools but it seems that they have a positive effect on students' attitudes towards biology in general and modern agriculture in particular. The implications of these findings are discussed in the light of modern approaches to inquiry learning.

Introduction

'*Students as researchers*' is a relatively new idea in education although "students' engagement in research activities is not uncommon in education. They often work on assignments and projects that require planning, collecting data/information, analyzing and writing of reports... students more often than not work on projects to obtain information which is already known and may be of little genuine interest to them or to the community." (p. 561, Atweh & Burton, 1995).

Schwartz (1988) who was amongst the very few researchers involved with projects, where students were employed as researchers asserted that: "by participating in a real inquiry, my student researcher assistants learned to think and behave like researchers" (p. 40). Two other projects where students participated as researchers were reported by Slee and Knight. Students were selected as researchers in these projects because of two reasons: "First, young people are well situated to identify and express their needs and further because, conducting the research provided an educational experience for the students that might be seen as worthwhile and valuable to their future employability" (p. 562, Atweh & Burton, 1995). Moreover, "students ...are clearly interested in and benefited by pursuing something beyond what they could do at school or at home" (p. 158, Kimbrough, 1995). All of the abovementioned instances state research relating to social sciences. However, the same approach, namely, 'students as researchers' was also adopted in the biological and the exact sciences.

An important task of science is to help students develop the thinking skills of scientists.

Whole programs, such as SAPA (Gagné, 1970) were designed to develop the reasoning skills scientists use in their investigations. The developers of the "new" science curricula, in the 1960's, appeared to be united in their conviction that introducing the laboratory as an integral part of the science curricula will enable students to internalize the spirit and method of scientific inquiry by providing them with opportunities to investigate, to inquire and to find things for themselves (Tamir & Lunetta, 1981). Yet, at present, inquiry is not taught effectively (Costenson & Lawson, 1986) and even in the laboratory, few students have the opportunity to develop thinking skills (Friedler, Nachmias & Linn, 1990). As a result, students often have the most rudimentary knowledge of science as a process and in many cases their concepts are erroneous and misled (Self et al. 1989). Schön (1983) described an example in which a group of Colombian high school students learned to experiment in open inquiry settings without specific instruction in science process skills. Such open inquiry, problem-oriented contexts are very desirable and commendable, according to him. In the same spirit, Roth (1993) conducted a study to investigate the development of science process skills in nontraditional laboratories where students engaged in open ended inquiry. The activity structures permitted students to pursue questions of their own interest within given content areas and to seek answers to these questions by planning and designing experiments, by collecting the data, and by transforming and interpreting the data. The findings of this study suggest that higher level cognitive learning was achieved by most students. This process of learner-based learning is self initiated, has a quality of personal involvement and is evaluated by the learner. The knowledge that evolves through this process is "private" knowledge, truth that has been personally appropriated (Rogers, 1969).

In the spirit of these educational approaches, Educational Greenhouses (EGs) have been used in Israel since 1979 and are currently installed in some 50 rural high schools. These facilities are attached to secondary schools and serve as a unique learning environment. Some of their specific features are:

- Enabling a much greater diversity of subjects and research designs (compared with the common Biology Laboratory)
- Affording true inquiry learning as students are actively involved in authentic research activity, throughout the project
- Execution of research projects using fully controlled research methodologies

Students' projects are often coordinated with work at regional or national R&D centers and supervised by scientists and specialists from research institutions. Within each EG, several projects are carried out simultaneously and depending on its physical size, up to 40 students may be involved. The projects form a compulsory component of the matriculation in Life Sciences for most students taking part. The research subjects span a wide range including for instance: acclimatization of wild indigenous flora, agro-ecology, pest management, zootechnics and aquaculture.

The interest in exploring this specific learning environment stemmed from preliminary observations conducted in an EG where the first author served as the principal teacher for Biology and related projects. Students attended the EG once a week over a period spanning a few months and added more hours if required. From the practice viewpoint they performed their tasks in a highly diligent manner: measuring, picking, tending their plants or animals. However, it was apparent that something was missing: students certainly had their "hands on" but was the project making them intellectually involved? In other words: 'hands on' = 'minds on'?

The present study was therefore designed to test the following issues:

1. Does involvement with inquiry related learning environments lead to a better understanding and interpretation of science process skills?
2. To what extent does the EG realize its potential as mentioned in the above features?

Methods
Research Tools

Since the main objective of this study was concerned with the understanding of learning processes in an hitherto unresearched learning environment, qualitative research methodology was selected as the main method of inquiry.

Different techniques including observations, interviews, a test and an analysis of the final written projects were undertaken in order to answer the first question. These techniques are described in the following:

Observations: Weekly non-participant observations were conducted by the principal author in two EGs and coincided with the students' research work. These observations consisted of recording verbal exchanges and behaviors of students, teachers and professionals associated with the research.

Interviews: One formal interview was conducted about 2 months after the initiation of each students' project. The interview consisted of questions which referred to the students' understanding of the relevance of their project and of their research plan. Many informal interviews were conducted throughout the research period, focusing mainly on the students' understanding of concepts and processes relating to their projects.

Test: This was administered following the completion of the practical component of the project (see Appendix 1). In view of the fact that most projects were associated with optimization of growing conditions of flowers' crop, testing the scientific skills of students was conducted using a simulated situation which called for the employment of similar thinking skills. In other words, we were interested in finding out their ability to transfer from one experimental situation to another.

The test consisted of two parts and was deliberately designed so as not to allow the use of "patterns" of inquiry thinking already well known and practiced, i.e. "formulas" used in the biology laboratory. In the first part, students were asked to refer to or comment on a simulated research situation. This simulation related to the effects of the following factors on annual potato crop: fertilization, watering, species, type of irrigation (sprinklers vs. drip) and temperature. In this part the skills tested were:

- formulation of a research question
- setting the hypothesis
- designing the experiment
- identification of dependent/independent variables

In the second part, students were presented with a set of experimental data which they were asked to analyze in terms of their ability to:

- make a valid judgment relating to:
  a. the correct interpretation
  b. the quality of the experimental design
- distinguish between experimental results and conclusions

Written projects: The final written projects which were submitted at the end of the research were analyzed by 3 independent referees (including the author). Validation of the analysis was assured by comparing the evaluation of the three referees. In this paper, attention is given to the interpretative skills of the students as demonstrated in the Discussion section in terms of:

– Explaining rather than reiterating the findings
– Drawing the correct conclusions from their findings

Additional information was gathered from informal conversations between the researcher and people who were associated with the EG. The main objective was to accumulate pertinent verbal information, types of interactions and typical behaviors, which would later serve for the setting up of a theoretical framework (Denzin & Lincoln, 1994).

The present paper relates specifically to findings based on the above test and preliminary findings drawn from analysis of the written projects.

To answer the second question, a survey of all final projects' proposals, submitted by the schools to the Education Ministry's Inspectorate of Secondary Schools, was undertaken. This survey relates to 1994, the year when this project was initiated.

Subjects

Two classes of seventh grade students, each from a different school, were chosen as samples of convenience. Since this research was 'case study' oriented, the specific selection of the two classes was not directly relevant. Class 1 had 28 students and class 2, 17 students of which 9 dropped out; these students decided not to continue with their project.

The study was scheduled as follows:

Class 1: from November 1996, when the students chose their projects' subject, to May 1998 when they submitted their written projects.
Class 2: from November 1996 to January 1998.

The reason for the difference in duration was due to the comprehensiveness of the two types of projects. In both cases, the students sat an oral examination, the examiner always being an outside person to the school where the projects were being held. Projects' marks were weighted in the final matriculation mark in Life Sciences.

Analysis of data

An Interpretive research methodology was adopted for the construction of meaning from the qualitative data. The test and the final written projects were analyzed, using criteria relating to each and every individual question or chapter.

Results

Preliminary findings derived from the Test

The following relate to the individual items mentioned in the Methods section:

1. *Formulation of a research problem*. The performance of almost all students was generally good. However, a problem has been identified with the definition of the dependent variable: some 46% of the students mistook the variable 'potato crop' for 'potato number', 'potato rate of growth' or 'potato price'.

The theoretical and practical aspects

Growth condtions	Right answer + explanation	Right answer – no explanation	Wrong answer	No answer
Fertilization	–	14.5	82	3.5
Amount of water	7	28.5	61	3.5
Species	3.5	18	71.5	7
Mode of irrigation	7	14.5	71.5	7
temperature	10.8	10.7	71.5	7

Table 1. *Distribution of answers relating to the effects that growth conditions may have on potato crop, as a percentage*

The ability to hypothesize. As evident from Table 1, 61% to 82% of the students did not answer correctly. Common misconceptions that were evident in student's answers were:

"A higher temperature is always superior to a lower temperature"
"The more fertilizers, the better"(for the crop)
The lack of use of proper scientific terminology was also very conspicuous:
"A normal amount of water (enough but not too much) will have a beneficial effect on the crop"
"A too low or too high temperature will result in less crop."

Experimental design. Approximately two thirds of the students were successful in designing a controlled experiment. Nevertheless, one third of the designs had more than one variable or lacked a control. Only 29% remembered to mention repetitions.

Identification of variables. Table 2 outlines the major findings. Common problems were:

   a. interchange of independent with dependent variable and vice versa.
   b. selection of all 5 growth conditions as a dependent variable

In the second part of the test, students were asked to judge the validity of a given (simulated) experimental design where no control of variables was specified, as well as relate to the conclusions drawn.

Table 3 summarizes the findings. Save one student, all students failed to

Dependent variable	Independent variable	
15	6	right answer
13	19	wrong answer

Note: 3 students did not refer to the independent variable.

Table 2. *Distribution of students' answers to skill tested: recognition of variables*

recognize the lack of control and went on to draw an invalid conclusion. An interesting finding was the fact that some students recognized and mentioned the presence of two independent variables but that did not deter them from drawing a conclusion. Another question that tested the ability to critically judge a bad experimental design showed that only 27% of the students were able to do so. Regarding the third item, students were given experimental data and interpretations based on the data. They were asked to consider 4 different interpretations and relate to them, giving their reasons for accepting / not accepting the interpretation. There were 63 right vs. 29 wrong answers; however, the reasons given for the specific choices were more often than not unrelated, irrelevant and/or incorrect. Finally, only 62% of the students were able to distinguish between results and conclusions.

A valid conclusion is possible		Valid conclusion is not possible	
+ explanation	no explanation	+ explanation	no explanation
10	13	1	0

note: 4 students did not refer to this question.

Table 3. *The ability to recognize an uncontrolled experimental design*

Preliminary findings derived from analysis of the final written projects

Agreement among referees was 82%. One of the most conspicuous elements was the absence of a true discussion of the results. In only 7 out of 26 final papers there was a genuine effort to deal in a scholarly way with the findings. It may be stated that most students mistook the Discussion for a summary of the results.

The theoretical and practical aspects

In addition, there was a clear dearth in terms of using statistical inference methods. Only 2 out of the 26 papers went as far as analyzing the data using statistical methods. Hence, although on the whole the findings, irrespective of the individual project, tended to come out in clear favour of one conclusion or the other, students were absolutely unaware of the meaning of certainty or confidence in scientific decision making. In about 50% of the projects correct conclusions were drawn and expressed more often than not in the form of recommendations to the growers.

Utilization of the educational Greenhouse

The cost of a middle sized, fully equipped EG, is estimated at around 100,000 – 150,000 US$. In addition, the salary of a full time maintenance person / coordinator has to be taken into account. Moreover, the teacher in charge is paid for the extra work she/he is doing. In some schools, a fully paid scientist from one of the research institutions will be invited to help design and lead the projects. This means that the cost of setting up and running an EG is very high indeed.

In a survey, carried out in 1995, two questions were posed. One related to the correspondence between the number of projects carried out in the EGs vs. their availability. As outlined in Table 4, not all of the projects are carried out in the EG. Moreover, although not apparent from this table, in 3 schools, EGs were not used at all. It seems that at the time of this survey, the EG was not considered the only learning environment suitable for research. The other question concerned the relative number of students involved in EG projects out of the total number submitting a project that year, as related to relative number of schools. The results are depicted in Table 5 which shows that only a small fraction of the school's students were involved in EG's related activity.

It is possible that certain shifts in the quoted percentages have occured during the last 4 years but in principle this is a valid description of the situation and one that indicates that the EG is still a long way from being a fully integrated environment for teaching and learning. Supported by the Education Ministry's Settlement Education Authority, there is now a growing awareness among teachers of the potential and sophistication of the EG with the consequence that these facilities are strongly recommended as learning environments for science.

% of projects carried out in the EGs	% of schools with EGs
1-25	25
26-50	42
51-75	25
76-100	8

Table 4. *Percent distribution of EG projects relative to their availability*

% of students involved in EG activity	% of schools with EG
up to 10	38
11-25	33
26-35	17
above 36	12

Table 5. *Percent distribution of students*

Discussion

An important task of science educators is to help students develop scientific thinking skills. Yet in spite of the costly and large efforts that began in the 1960s, inquiry is not taught effectively (Barufaldi & Swift, 1980; Costenson & Lawson, 1986). Furthermore, even in the laboratory, few students have the opportunity to develop these thinking skills (Friedler et al. 1990; Novak, 1988).

One of the main objectives of this study was to examine, in students, the relationship between supervised involvement in true authentic research and the development of scientific thinking skills using the Educational Greenhouse as the platform. Various studies on performance in specific domains have indicated that expertise depends on a collection of schemata specific to the domain (Friedler et al., 1990). Schön (1983) concluded that content and skills are so intricately interrelated that means become interdependent with ends, knowing becomes inseparable from doing. There is evidence to suggest that our ability to control and orchestrate cognitive skills consists of cognitive activity tied specifically to context (Greeno, 1988). It should also be added, practical problem solving proceeds through the use of contextual cues that interface with tacit knowledge rather than through the systematic application of explicit predesigned steps in problem solving (Rogoff, 1984;

The theoretical and practical aspects

Schön, 1987). However, efficient problem solvers apply these mental representations metaphorically to new situations.

Since important aspects of cognitive activities are functions of meaningful contexts, an assumption was made, that students' activities in the EG could enhance their understanding and interpretation of inquiry thinking skills. The findings do not support this hypothesis as evidenced by a mediocre mastery of the various process skills which were examined. It seems that in spite of the specific features and unique opportunities in this learning environment, the process of enculturation into science thinking has not fully materialized. Students' identification of variables as well as interpretation of experimental results was lacking in spite of the familiarity with the physical and conceptual context. Moreover, although the exposure to a real environment and the experience gained through it may have added to their knowledge, it is highly doubtful whether the students benefitted in terms of acquiring a new way of looking and interpreting of natural phenomena.

Though this study did not include assessment of the students' aptitude in general, it is contended that they did not deviate materially from the average. In other words, the somewhat frustrating results cannot be attributed to a general low achievement level. Rather, problems of implementation, including the absence of a model for teaching in the Educational Greenhouse, may be cited as one of the major factors underlying its limited success. It is therefore suggested that there is an urgent need for the implementation of such a model which will incorporate aspects from various relevant learning theories in science education: discovery learning, experiential learning and constructivism. We also recommend that elements of statistical decision are taught in conjunction with the study of biology and interpreted in the light of normal variations in living organisms.

As for the realization of the EG's potential in terms of the abovementioned specific features, it would seem that the EG is at least partly fulfilling its expectations. In schools with EGs, throughout the country, many students are involved in true, authentic research spanning the fields of Biology and Agricultural studies, greenhouse structure engineering, plasticulture, computer programming, electronic control equipment and economics. Additional subjects such as poultry, fish and aquaculture have also been introduced recently to the greenhouse learning activities. Increasingly, EGs are moving from vegetable and houseplant research to new fields like zootechnics, fruit and woody plants and agro-ecology (including biological pest control). Furthermore, thanks to the program, youngsters who previously thought of agriculture as just hard menial work, are learning to appreciate it as advanced scientific activity.

References

Atweh, B. & Burton, L. (1995). Students as Researchers: rationale and critique. *British Educational Research Journal*, 21 (5), 561-575.

Barufaldi, J.P., & Swift, J.W. (1980). The influence of BSCS – Elementary school science program instruction on first-grade students' listening skills. *Journal of Research in Science Teaching*, 17, 485-490

Costenson, L.J., & Lawson, A.E. (1986). Why isn't inquiry used in more classrooms. *The American Biology Teacher*, 48, 150-158

Friedler, Y., Nachmias, R.. & Linn, M. C. (1990). Learning scientific reasoning skills in microcomputer-based laboratories. *Journal of Research in Science Teaching*, 27, 173-191

Gagne', R.M. (1970). *The conditions of learning* (2nd ed.). New York: Holt Rinehart and Winston.

Greeno, J.G. (1988). *Situated activities of learning and knowing in mathematics*. Paper presented at the 1988 annual meeting of the PME-NA, DeKalb, Illinois

Kimbrough, D. (1995). Project design factors that affect student perceptions of the success of a science research project. *Journal of Research in Science Teaching*, 32 (2), 157-175.

Novak, J.D. (1988). Learning science and the science of learning. *Studies in Science Education*, 15, 77-101

Rogers, C. (1969). *Freedom to learn*. Columbus, OH: Merrill.

Rogoff, B. (1984). Thinking and learning in social context. In: B. Rogoff & J. Lave (Eds.), *Everyday cognition: Its development in social context*. Cambridge, MA, Harvard University Press, pp, 1-8

Romberg, T.A. & Carpenter, T.P. (1986). Research on teaching and learning in mathematics: two disciplines of scientific inquiry. In: M.C. Wittrock (Ed.). *Handbook of research on teaching*. New York, MacMillan, pp. 850-873

Roth, W. M. & Roychoudhury, A. (1993). The development of science process skills in authentic contexts. *Journal of Research in Science Teaching*, 30 (2), 127-152.

Schön, D.A. (1987). *Educating the reflective practitioner*. San Francisco, Jossey-Bass, pp. 56-78

Schön, D.A. (1983). *The Reflective Practitioner: How professionals think in action*. New York: Basic Books.

Schwartz, J. (1988). The drudgery and the discovery: Students as research partners. *English Journal*, 77, 37-40.

Self,C.C., Self, M.A. & Self, D.C. (1989). Science as a process: Modus Operandi. *The American Biology Teacher*, **51** (3),159-161.

Strauss, A, Corbin, J. (1994). Grounded theory methodology. In N.K. Denzin & Y.S. Lincoln (Eds). *Handbook of qualitative research*. Thousand Oaks, Sage Publications, pp.273-285

Tamir, P. & Lunneta, V.N. (1981). Inquiry- related tasks in high school science laboratory handbooks. *Science Education*, **65** (5), 477-485.

Miki Dvir and David Chen

Appendix 1

Test: The Case of Potato Crop

"In a kibbutz which grows potatoes, a decline in crop has been noted. To remedy the situation, an agronomist was recruited whose function it was to advise on optimization of growth conditions for the achievement of maximal crop.

The agronomist performed a series of controlled experiments in order to study the influence of the following parameters on the crop:

1. Fertilization (addition of minerals namely P, N, K as compared to no added minerals)
2. Watering (100, 120, 160 m^3)
3. Potato's species ("Alpha" or "up to date")
4. Type of irrigation (sprinklers or drip)
5. Temperature (15 or 25° C)

The experiments were conducted on a plot measuring approximately one quarter of an acre. The crop in each and every plot was measured and recorded".

In all, the test consisted of 23 questions which were grouped into 7 headings. In the following, a sample of questions relating to the above case is presented.

Please answer the following questions:
1. What is the research question?
2. How would the parameters: Fertilization, watering, species, type of irrigation, temperature, influence potatoes' crop? Put in hypothesis form.
3. The agronomist would like your advice concerning the research plan. Please choose one parameter and draw a research plan. Remember, you want to be able to tell how this parameter would influence the crop.
   Specify the independent and dependent variables in your research plan.
4. Potato crop was examined under the following conditions:
   no fertilization, 15°C, watering – 120 m^3, irrigation by sprinklers, "alpha" species – crop was 240 kg.

b. no fertilization, 25°C, watering – 120 m³, drip irrigation, "alpha" species – crop was 220 kg.

What conclusion can you draw regarding the conditions influencing potato crop? Please explain your answer.

It was found that potato crop grown at 25°C was larger compared to potatoes grown at 15°C. This statement may be considered as a:

a. Research hypothesis
b. Research result
c. Research conclusion

Changing teachers' practice
Practical work in environmental education

Peter van Marion

Teachers may hold different views on the meaning of fieldwork in education. The views range from seeing fieldwork simply as a tool in science education, through seeing fieldwork as an opportunity to give pupils experiences of nature, to approaching fieldwork as activity closely related to environmental management in the local community. Some of the views on fieldwork harmonize well with current trends in environmental education.

The concept of a national environmental education network programme in Norway is presented. An evaluation of the development of a part of the programme, shows that different views on the nature of fieldwork in environmental education are held by the persons involved in the development of the programme. These different points of view indicate which hindrances the implementation of the programme may meet, both inside and outside schools.

Introduction

This paper deals with the role practical work in nature study can play as an element of environmental education. Possible reasons, whether conscious or subconscious, for the choice of content and form in practical work will be discussed. How central school authorities in Norway try to facilitate practical work outside the school being aimed at those goals set for environmental education will be illuminated. Further, it will be suggested how one can expect teachers and others to perceive and react to this approach to practical work.

Environmental education is not a separate subject in Norwegian schools.

The official strategy for environmental education in schools aims at integrating it into every subject. Practical work connected to environmental themes must therefore either take place within the framework of traditional school subjects or within the framework of special interdisciplinary environmental projects. Considering the close connection between science education and environmental education, it is not surprising that many teachers choose to make practical nature study a central element when attempting to achieve environmental goals.

Going on an excursion with a school class is a demanding experience. There is no other type of teaching which makes such vast and numerous claims on the teacher. He or she has to be a capable organiser and administrator, be able to function as a mechanic and first-aider and be a well informed and understanding educator. Last, but not least, the teacher must have the scientific insight essential to the specialist work involved. Despite all these exacting challenges and a tremendous work load, many teachers plan and carry out fieldwork with great enthusiasm and drive. It is obvious that these teachers are driven by a conviction that fieldwork ought to have its place in the instruction offered by schools. This conviction must build on some idea about how fieldwork can contribute to reaching the goals one has for one's teaching.

It seems reasonable to assume that most teachers are of the opinion that fieldwork is a method which has special educational qualities. This does not seem to be a particularly modern viewpoint. Ford (1981) shows that even Rousseau (1712-1778) and Pestalozzi (1746-1827) were fascinated by the fact that children learn best through encounters with natural phenomena out in the countryside. Dennis & Knapp (1997) show that Dewey (1859-1952) emphasized that 'nature study' ought to be just that. Through direct observation pupils would, according to Dewey, be able to gain insight into interactions in nature and between nature and society.

In more recent literature (see e.g. Haugset et al. 1979; Kvam et al. 1989) a number of possible reasons for incorporating fieldwork into science education are pointed out. These can be summarized thus:

1. Subject matter becomes tangible.
2. Subject matter is placed in realistic contexts.
3. The pupils gain insight into the practical application of scientific knowledge.
4. Pupils gain experience in the use of scientific methodology.

5. Pupils gain experience in the use of scientific instruments.
6. Pupils' ability to observe is developed.
7. Pupils' ability to cooperate is cultivated.
8. Pupils develop positive attitudes towards outdoor life and the utilisation of nature.
9. Pupils gain insight into the connection between nature and the society outside the school walls.
10. Pupils gain a fuller understanding of, and interest in, a justifiable use of resources both locally and on a global scale.

Several of these arguments are not only valid for science teaching, they can also apply to the use of fieldwork in education in a wider sense, for example in environmental education. Earlier this century Dewey saw clearly that nature study has an interdisciplinary aspect and there should be a close connection between science education and environmental education (Dennis & Knapp 1997). He also pointed out the importance of fieldwork in environmental education. Today there is general agreement on the matter. Even though it is clear that environmental education differs from science education on a number of important points it seems natural that scientific fieldwork constitutes a central element in environmental education in schools.

Fieldwork and its organization

Traditionally fieldwork in science education has often been regarded as a chance to illustrate textbook theory (Thomas 1990). On a classical excursion the teacher demonstrates and guides the pupils around in nature. Nature is used to prove that the theory is correct.

Alternatively, the pupils are instructed to, themselves, carry out investigations in nature. Although the pupils seem to play a more active role during investigative excursions, the excursions still very often aim mainly at confirming textbook theory. Form and methods in more investigative excursions differ from the classical excursion, but the main objectives are not necessarily different.

Regardless from form and methods, fieldwork may, however, aim at more than showing that textbook theory is correct. Whether or not this is recognized by teachers depends on their views on the main aims of fieldwork and how it should be organized. The views held by teachers may vary widely and are closely related to their understanding of the meaning of the study of the natural environment.

Different views on science education and fieldwork

Östman (1996) analysed Swedish science textbooks and was able to distinguish between three main views on science education.

Disiplinary science education can be characterized as aimed mainly at pupils learning scientific concepts, theories etc. In so far as nature is used, it is in order to illustrate and concretise abstractions. Nature is merely an aid to learning.

In practical-applied science education weight is put on mankind's vital necessities, the advantages and importance of nature for human beings and our utilization of nature and the consequences of this. Pupils are to learn how scientific knowledge can be used to understand and thereby control our necessities, activities and their effects.

In moral-applied science education the emphasis is on pupils learning about human activity and its environmental consequences using their scientific knowledge. In addition pupils are to learn how we should conduct ourselves with regard to nature and what our environmental moral responsibilities are.

Different views on science education, as Östman describes them, will also be expressed in different opinions on the importance of practical nature studies, different ways of organizing these and different reasons for one's choice of content and methods.

Even though it is the classical excursion with the teacher as an all-knowing guide which is most commonly connected to Östman's disiplinary science education, a more investigatory fieldwork can also, in principle, take place within the framework of disiplinary science education. On the other hand both types of excursion may also be consistent with Östman's practical-applied science education. It is, however, more difficult to imagine that the classical excursion can be an appropriate way of organizing fieldwork when emphasis is on moral aspects of the interaction between nature and man. It will then be more natural to base fieldwork on personal investigative activity and greater freedom for pupils to organize their work.

Peter van Marion

Participation and collaboration

In 1992 the United Nations summit meeting in Rio de Janeiro agreed upon a plan of action, Agenda 21, aimed at ensuring an economically, socially and ecologically sustainable development. Agenda 21 gives weight to the role children and young people must play in protecting the environment and their participation in decisions about environment and development. The Rio-conference also recommended development of local Agenda 21s. Through local Agenda 21 work the interests of children and young people may be brought into planning processes and other decisions about the local environment.

Participation in local environmental work may lead to active involvement in the local community. This provides a new learning context which motivates the pupils and contributes to the development of pupils' awareness of a wide variety of aspects of the issues studied. These include not only science related aspects, but also social and economic aspects. Carrying out fieldwork in this context helps the pupils to see their investigations as something meaningful with relevance to environmental management, instead of being just 'exercises' in science education.

Participating in local environmental work involves collaboration with persons and institutions outside the school. Formerly collaboration between schools and institutions outside the school was simple and usually directed in only one way. Schools received support from outside institutions through practical help from experts, printed information, excursions to the institution etc. A new pattern of cooperation between schools and the community shows that schools may become a resource for local authorities or researchers in their environmental work. Schools and other institutions become 'partners' in environmental work. Some school classes provide their outside partners with environmental data, others carry out environmental actions. Examples are monitoring environmental conditions in a nearby stream or lake, registering discharges of waste, working out plans for sorting waste or the use of nature areas and restoring the local habitats of endangered species.

Based upon these ideas, local, national and international programmes have been established in order to stimulate schools to participate in local environmental work. The programmes GREEN and GLOBE are well-known examples, and several other programmes have been established in many countries (see e.g. Ewing 1990; Monroe & Wals 1990; Suti 1991; Dvornich,

Tudor & Grue 1995) In most of these programmes schools are offered practical and methodological help in order to carry out sampling and registration in the local environment, the results of which may be entered into databases which are open for researchers and environmental managements.

Making decisions about fieldwork

When deciding about the content and form of planned fieldwork, teachers and pupils have to make many decisions. Important matters to be decided are mentioned above. Summarizing, it may be stated that the main decisions to be taken deal with to what degree the fieldwork should be based upon:

- studying the relationship between nature and society
- freedom when choosing working methods
- participation in local environmental work through collaboration with agencies outside the school

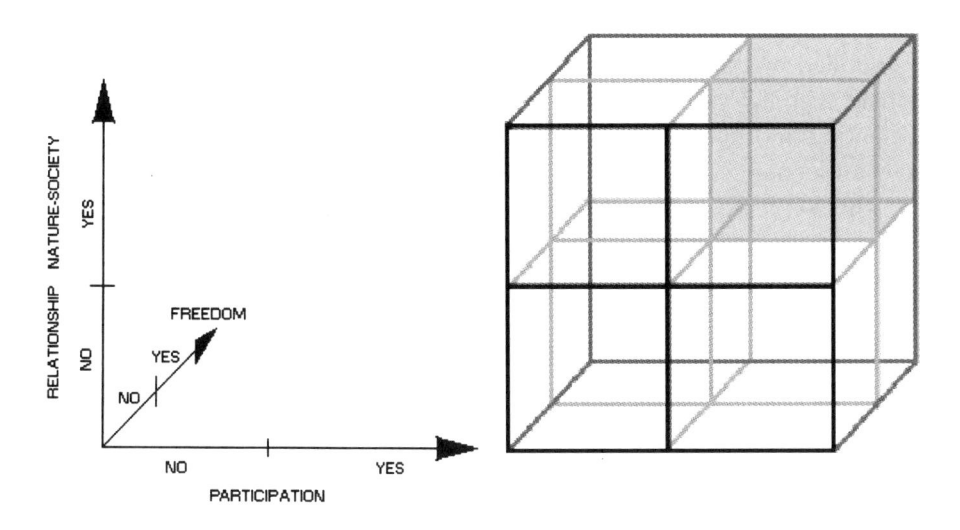

The range of possible combinations that different decisions may result in, are visualized in a three-dimensional model, in which each of the axes represents one of the three main decisions to be taken. In order to simplify things, the model is based upon only two values of each one the three variables, yes and no. The model results in eight types of fieldwork, all based on a different set of preconsiderations. In reality the range of various types of fieldwork may, of course, be larger, since each of the variables may have many, rather than only two values.

The Environmental Education Network in Norway

Norwegian educational authorities have been making efforts for several years to provide schools with tools for creating action-oriented education (Sandås & Benedict, 1993). Several national programme, aiming at supporting schools in environmental education were developed and offered the schools by educational authorities, NGOs and others. For practical and economic reasons, it was not desirable to continue a large number of environmental programmes. In 1997 most of these programmes were therefore restructured and organized as one programme, which is called the Environmental Education Network (Ministry of Education, Research and Church Affairs et al. 1997).

Environmental education involves challenges in both subject matter and teaching methods which it may be difficult for schools to solve satisfactorily. By creating a system for contact between schools, the environmental authorities, and research institutions, it is possible to spread information and establish routines that will give the education sector access to updated environmental information as well as opportunities to develop knowledge through active participation.

In the Environmental Education Network, schools are offered suggestions for practical fieldwork related to environmental issues, and quality-controlled methods for collecting environmental information are available. There will always be a competency centre available to schools to answer scientific and methodological questions that will arise when working with environmental issues.

The Network enables schools to carry out tasks that are useful to society in cooperation with experts outside the school. This will make it possible to exchange information in both directions between schools and society. Suggestions are made as to what kinds of environmental improvements could be made, with guides for how various kinds of activities can be carried out. By following these methods and guides, the pupils' efforts to learn can, at the same time, be useful for the local authorities.

Restructuring and further development of the programmes in the Environmental Education Network

The restructuring of the existing programmes into a network may make environmental education more widespread in schools, particularly since computer technology has now become more widespread in schools. The restructuring process also includes further development of some of the programmes. Until now, three existing programmes related to water habitats have been revised and developed further within the concept of the Network towards one water programme.

The development of the new water programme was initiated in 1996 and was expected to be completed in 1997. The Ministry of Education, Research and Church Affairs also wished the development of the new water programme to include information which could form a basis for the development of an implementation strategy for the concept 'Environmental Education Network'. In accordance with this an evaluation of the development of the water programme was carried out (Knutsen & van Marion 1997).

The Evaluation of the water programme
Organization of the development of the water programme

The responsibility for the development of the new programmes was placed on the National Centre for Educational Resources. The development of suggestions for fieldwork activities was planned as a project in close cooperation with the County Offices of Education in the counties Oppland and Telemark. In each of the counties the County Offices of Education put together a working party composed of teachers at all levels, a representative from the County Office of Education and representative(s) from environmental management. The working parties main task was, first and foremost, to generate suggestions for activities through adaption of parts of the existing programmes, production and testing.

The intention of the chosen development strategy was to optimise the process through

- making use of the expertise which teachers/schools represent on environmental training and development at school and class level.
- making use of the expertise local environmental management units rep-

resent on environmental management questions, interdiciplinary collaboration, knowledge of the vertical system within environmental management, as well as potential challenges for schools in the local environment.
- making use of the expertise school owners and County Education Offices represent in e.g. coordination and organization, school development, initiatives with environmental education, competency development and interdisciplinary collaboration at local authority and county level.
- making use of the advantage of having participating teachers from the same school.
- making use of the potential which lies in the work of development within fairly small groups.

Evaluation of the process

The evaluation can be characterised as a participatory evaluation; through an active role in the process and continuous communication with all the persons who were involved, a documentation basis for the evaluation was obtained.

We took an active part in all the meetings in the working parties, worked out reports after the meetings and asked participants to give their comments. The participating teachers were asked to answer questionnaires and fill in report forms during different phases of the process. In addition structured individual interviews were carried out with all participants close to the end of the process, in order to obtain information about the individual overall viewpoints about the process and its results.

Furthermore, we evaluated the drafts of the suggestions for activities, as they were produced en route by the teachers. The drafts were categorized as to which degree they supported different roles of the school in collaboration with society:

1. The school as a receiver of information and/or technical, methodical and practical help.
2. The school as a receiver of information and/or technical, methodical and practical help. At the same time, the school reports in such a way that partners outside the school can make use of the school's work. The outside partners may help in setting out problems and propose concrete tasks for school classes.
3. The school as a receiver of information and/or technical, methodical and

practical help. At the same time the school contributes to environmental work in the local community, or the school supplies premises for, and takes part in the local environmental debate.

It is mainly forms of cooperation such as 2 and 3 which contribute to developing the school as a socially active school. Corresponding to this, suggestions for fieldwork activities which could be placed in categories 2 and 3, could be considered as being in correspondence with the concept of the environmental education network

Results and discussion

The total impression was that a relatively large number of the drafted activities, in so far as they were aimed at co-operation with participants outside schools, were mainly aimed at collaboration as described above in category 1. This indicates that some of the teachers who took part in this process did not quite manage to put the fundamental concept of the project into practice. This was surprising, considering the emphasis placed on communicating the concept in the project groups. This may, in part, be due to the framework for the task, for example: the deadline, a feeling of uncertainty at the start of the project and certain organisational aspects. Even so the results indicate that there must also have been insufficient understanding of the fundamental ideas or some of the participants must have difficulty in fully accepting these ideas. Discussions en route may indicate the latter.

The discussions amongst those participating in the developmental process show different points of view on a number of important factors. A joint understanding of several of these factors was reached but in some areas the dissension was fundamental and the differences continued to exist.

The points of view which seemed to provoke major dissension had to do with the significance and purpose of undertaking the collection and registration of data based on requirements defined by research institutions. Different viewpoints about the type and quality of the data and the need for quality control appeared to be a consequence of this difference of opinion. On the whole the discord here seemed to follow school levels. Teachers from senior secondary and, to a certain extent, junior secondary schools were more disposed to see the relevance of collecting and registering physical, chemical and biological parameters in collaboration with scientists than primary teachers. This was no great surprise, it implies that the different levels, for natural reasons, have different needs and make different demands on a national environmental education network.

Among participants from municipal and county administration there were also differing opinions about what to emphasise and to what extent there was a need for data which schools could obtain. There was, however, agreement that management, particularly at the local level, would benefit from close co-operation with schools where utilisation and planning of areas was concerned. The local Agenda 21 work was seen as a joint arena for management and schools.

One subject, which was discussed in different contexts, was how to interpret management's need and motives for co-operating with schools. Is it first and foremost the aim of environmental management to contribute to putting environmental topics on the classroom agenda, hopefully leading to greater environmental commitment and, in this way, contribute to the promotion of positive attitudes in the population at large? In this way schools would be an obedient tool for a management apparatus whose goal is to contribute to the development of positive attitudes. Or does management wish a more pro-active school, which through activities in collaboration with management will, in certain cases, have a genuine influence, directly or indirectly, on the administrative level? If that is the case more critical pupil attitudes may arise and there is a risk that schools, in some cases, will criticise the authorities.

Different opinions were also put forward in discussions about to what extent it is proper to let pupils carry out 'tasks' for researchers. Should the requirements of management and research be the main basis for practical activities – or should the choice of activities be primarily based on educational evaluations? Should the aims of the curriculum be the main reason for the choice of subject and the basis of practical activities – or should the starting point be national environmental goals?

The participants' evaluation has provided valuable information on different aspects of the process and its supervision, which will be of use in the subsequent phases of the development of the environmental education network. Of particular importance is the fact that the evaluation has given insight into some of the barriers which must be overcome if the implementation of the network is to be successful. The opinions of the participants put forward in discussions seem to indicate which problems may be expected to emerge amongst the programme's users, both in schools and amongst the school's collaborators.

Change in teaching practice?

Fullan (1991) points out that 'educational change depends on what teachers do and think – it's as simple and complex as that' (p.117). The environmental education network represents an exciting rethinking on the role practical work can play in both science education and environmental education. For teachers the concept represents a genuine ideological alternative to the different traditional theories on the use of fieldwork. Moreover the network represents an educational tool which is practically organised in such a way as to give schools a low use threshold. If it will be possible to alter teaching practice on practical work, or if teachers will continue on the same track, depends to a large extent on whether teachers are able to perceive the educational dividend which is the notion behind the environmental education network.

References

Dennis, L. J. & Knapp, D. (1997). John Dewey as Environmental Educator. *The Journal of Environmental Education*, 28(2), 5-9

Dvornich, K. M., Tudor, M & Grue, C. E. (1995). Nature Mapping: Improving Management og Natural Resources through Public Education and Participation. *Wildlife Society Bulletin*, 23(4): 609-614

Ewig, S. (1990). Saltwatch – Involve me and I'll Understand.

In J. Van Trommel (ed), *Proceedings of the International Symposium on Fieldwork in the Sciences*. SLO. Enschede, pp. 123-132

Ford, P. M. (1981). *Principals and Practices of Outdoor Environmental Education*, John Wiley and Sons, N.Y.

Fullan, M. G. (1991). *The New Meaning of Educational Change*. Teachers College, Columbia University

Haugset, O., Hjelmfoss, P., Langangen, A. & Litland, G. (1979). *Metodisk veiledning i biologi*. Forsøksrådet for skoleverket, Rådet for videregående opplæring, Gyldendal Norsk Forlag

Knutsen, A. E. & Marion, P. van. (1997). The National Environmental Programme for Actionoriented and Interdisciplinary Environmental Training in the School's Own Area; An Evaluation of the Development of the Subprogramme Water. *PS-report 1-97*. Program for Educational Research. The Norwegian University of Science and Technology

Kvam, A., Størkersen, I.O. & Valdermo, O. (1989). *Metodisk veiledning i naturfag*. Rådet for videregående opplæring, Gyldendal Norsk Forlag

Peter van Marion

Ministry of Education, Research and Church Affairs (1995). *Nationwide Environmental Education Programmes.* Evaluation Report. KUF, Oslo

Ministry of Education, Research and Church Affairs & Ministry of Environment (1997). *Environmental Education Network.* KUF, Oslo

Monroe, M.C. &. Wals, A. E. J. (1990). From Fieldwork to Action: Interactive River Studies. In J. van Trommel (ed), *Proceedings of the International Symposium on Fieldwork in the Sciences.* SLO. Enschede, pp. 103-114

Sandås, A. & Benedict, F. (1993). *Environment and School Initiatives; Indepth Policy Review in Norway.* Background Report for OECD Expert Team. KUF, Oslo

Sutti, S. (1991). The Water Analysis Project (WAP): An Alternative Model for Environmental Study. In *Environment, Schools and Active Learning.* Centre for Educational Research and Innovation, Paris

Thomas, T. (1990). World Trends in Fieldwork. In: Trommel, J. van (ed.): Proceedings of the International Symposium on Fieldwork in the Sciences. SLO. Enschede, pp. 15-21

Östman, L. (1996). No-dikatiska perspektiv på undervisning och i lärerutbildning: en artikkelserie om meningsskapende målrealisering och lärerkunskap. In O. Eskilsson & G. Helldén (eds), *Naturvetenskapen i skolan inför 2000-tallet.* Fagus, Högskolan Kristianstad, pp. 552-595

Section 5

Models of student cognition in practical work: Perspectives from one research programme

Introduction

A group of researchers at the University of Bremen have worked for some years on a theoretical framework based on 'situated cognition in time', under the leadership of Stefan von Aufschnaiter. This in turn has allowed them to develop methodologies for investigating students' learning processes. The three papers in this section are examples of research where students together with a tutor are performing activities in a special laboratory setting. The learning processes of the students are recorded and interpreted and the results highlight some of the important aspects of learning in a physics laboratory.

The first paper, by Claudia von Aufschnaiter, Anja Schoster and Stefan von Aufschnaiter, reports an investigation of the interdependency of the learning environment and students' learning, and stresses the significance of the affective as well as the cognitive domain for learning. The results of an extensive use of a questionnaire shows that a rewarding learning environment is dependent on the interest and level of cognitive development of the individual student. Learning and the involvement in teamwork with other students then depends on the extent to which the individual student finds the learning environment matches with students' own interests and own level of cognition.

The next paper by Anja Schoster and Stefan von Aufschnaiter also investigates in some detail students' cognitive processes in relation to the learning environment. The results of careful analysis of videotapes demonstrate how meaningful learning and the development of concepts depends upon students' situated pre-knowledge. In the process of learning, the route from less complex to more complex conceptions is paved with short time intermediate situated knowledge The teacher with the more comprehensive knowledge may be tempted to make shortcuts which can be harmful to learning.

The last paper by Manuela Welzel, Claudia von Aufschnaiter and Anja Schoster uses videotaped records of students' activities to investigate how time-scale, context and complexity of conceptions influence students' cognitive processes. Following this analysis, recommendations are given for teachers as to how to handle each of these instances in teaching situations.

Albert Paulsen

The influence of students' individual experiences of physics learning environments on cognitive processes

Claudia von Aufschnaiter, Anja Schoster & Stefan von Aufschnaiter

This investigation analyses how students' individual experiences, such as enjoyment and feeling of competence, interrelate with their cognitive development. For this purpose students' actions within learning environments were videodocumented. Using further questionnaires we asked students about their reactions to the learning situation they had just experienced. Analysing the videos by reconstructing students' situated cognitions leads to a detailed description of students' cognitive development. Analysing the questionnaire and comparing the video data and questionnaire data leads to the identification of interrelations between cognitive development and individual experiences. The results show that individual experiences are on short time scales (around five minutes) and are related to the results achieved within this period (in the students' view).

Introduction

Many approaches in pedagogical and educational research focus on motivation and interest as important factors influencing learning processes (e.g. Renninger, Hidi & Krapp 1992; Krapp 1996). Deci and Ryan emphasise the importance of students' feelings of autonomy, competence and their experience of the social surroundings during actions within learning environments for their intrinsic motivation (e.g. Deci 1992; Deci 1995; Deci & Ryan 1993). It is assumed that intrinsically motivated students learn 'more deeply' and achieve better results than extrinsically motivated students (e.g. Wild,

Schiefele & Krapp 1995; Schiefele & Schreyer 1994). Although it seems clear that students need positive experiences and interest for 'effective' learning processes to take place, little is known about the nature of the interrelation between these experiences and students' cognitive development. In this paper we will present some of our recent results about students' situated cognitions and individual experiences in physics learning environments.

Situated cognition

While Brown, Collins & Duguid (1989) and Lave (1997) describe situated cognition as contextual and socially related cognitive processes, for us the term 'situated' primarily means situated in time. From a neurobiological point of view the cognitive dynamic must be described as sequences of 'Images-of-now' (Damasio 1994). Each Image-of-now has to be 'produced' by the cognitive system (neuronal structure) within three seconds (Pöppel 1994). For experiences of a much larger time than three seconds, the cognitive system will produce sequences of Images-of-now which are connected to a greater or lesser extent, among other things, with changes in the learning environment. Each Image-of-now depends on the capacity of the working memory and therefore deals with only a small part of the environment and those recollections which are possible within three seconds. Cognition means all elements of perception, recollection, expectation and action which are active (existing neural networks are firing) and relevant on the same time scale, that is within the three seconds of each Image-of-now (Clancey 1993). Only within three seconds can all these cognitive activities be bound together into one Image-of-now (Pöppel 1994) by the cognitive system.

Learning is also situated in time but on time scales of sequences of Images-of-now. Learning is correlated with 'the success' of sequences of Images-of-now when solving problems and leads to changes in the cognitive system (neuronal structure).

Previous research work by our group concerning the learning processes of students (lower and upper secondary as well as university students) lead to the detailed analysis and description of learners' situated cognitions on time scales of about three seconds. For this purpose we videodocumented students' (verbal) activities in (experimental) learning environments and reconstructed individual situated cognitions from their actions (Fischer & von Aufschnaiter, S. 1993; von Aufschnaiter, S. & Welzel 1996, 1997a; Welzel 1997a/b).

Research questions

While gathering data and analysing them we did not yet focus on learners' individual experiences during actions in the learning environment. However, emotion and motivation are an essential part of situated cognitions (e.g. Damasio 1994). To find out whether students experience feelings like enjoyment, competence, and interest, several questions should be answered:

- Do learners enjoy actions within the learning environment?
- How do learners think about their own competencies concerning the learning environment?
- Do learners introduce their own ideas into the learning environment?
- Is teamwork important?
- Do learners consider the results they achieve to be interesting?

There may be a lot more questions concerning students' individual experiences (e.g. Csikszentmihalyi & Larson 1987), but for us these seem to be the interesting ones. With reference to these questions we developed a short questionnaire (see figure 1) to be completed by learners, concerning specific elements of the learning environment. The students' answers will be connected to the analysis of the videodata to investigate the influence of students' individual experiences of physics learning environments on their cognitive processes.

The following section presents the context of one of our recent studies and explains how the questionnaire shown in figure 1 was integrated.

Context of study

In summer 1997 we asked 30 students of grade 11 (age 16-17) to come to our university for three labwork sessions (each lasting about 90 minutes) in the field of electrostatics, and for two additional sessions of concept mapping and interviews (Schoster & von Aufschnaiter, S. 1998, and in this sec-

Claudia von Aufschnaiter, Anja Schoster & Stefan von Aufschnaiter

? interesting – important – difficult ?

Name: _____

No. of last card you worked on: _____

	not true	partly true	mostly true	completely true
I enjoyed carrying out the experiment.				
I have the feeling that I knew all that was needed to carry out the experiment successfully.				
While doing the experiment I was able to include my own ideas.				
Teamwork with my partner students was important for arriving at the results.				
I find the result interesting.				

What I really did not like about the experiment:
_____
_____

– What I liked most about the experiment:
_____
_____

VC:	video camera
MB:	box with material for electrostatic experiments
S1:	student 1
S2:	student 2
S3:	student 3
T:	teacher
Ta:	table
B:	blackboard
D:	door
W:	window

Figure 1: *Questionnaire on students' experiences*

tion). Most of these students (24) had decided not to choose physics in their last three years at school because they were not interested in the subject, considered the subject to be too difficult, or chose other science subjects (biology and/or chemistry). We videodocumented the laboratory and the additional sessions.

The influence of students' individual experiences

```
VC:  video camera
MB:  box with material
     for electrostatic
     experiments
S1:  student 1
S2:  student 2
S3:  student 3
T:   teacher
Ta:  table
B:   blackboard
D:   door
W:   window
```

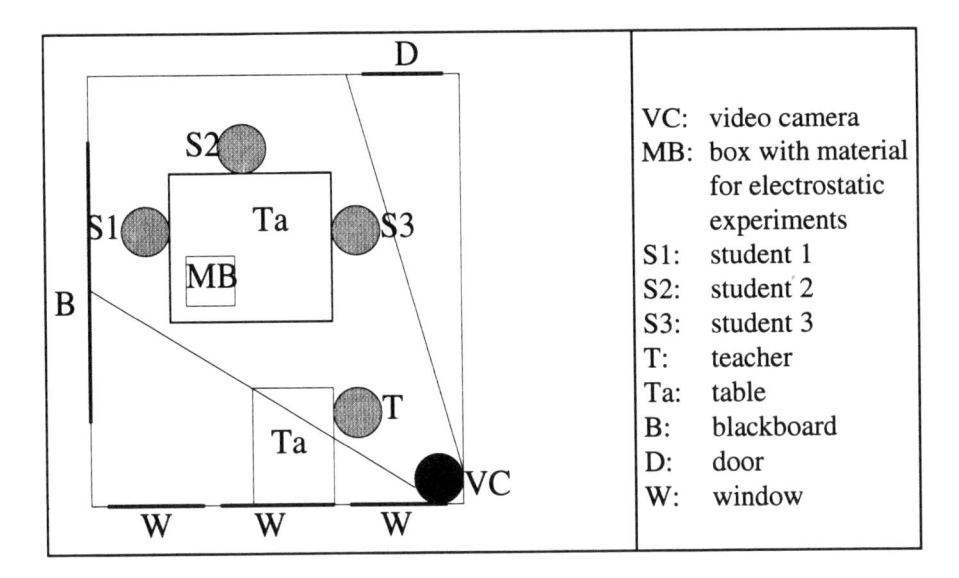

Figure 2: *Setting of the laboratory sessions*

During the laboratory sessions the students sat around a table in groups of three. They were given a box of materials (see figure 2) and cards with tasks, mostly connected with experiments, written on them (for example, see figure 3). Altogether there were 51 cards numbered from 1.01-1.21 (first session), 2.01-2.14 (second session) and 3.01-3.16 (third session). In addition to these cards the students received intervention cards and cards with thought experiments at specific points from the teacher. Some cards authorised the students to ask for information (also written on cards) from the teacher (for an example see figure 4).

What does it mean, that sometimes one end and sometimes the other end of the neon bulb flashes?

If you need help, you can ask the teacher for more information.

We did not do this card because

Figure 3: *Card 2.11*

285

> **Explanation of the neon bulb**
>
> Using a neon bulb you can distinguish whether an object is positively or negatively charged. The neon bulb flashes at the end which touches the negative part of the object.
>
> Do you think that this information can help you in solving future tasks?
>
> ☐ Yes   ☐ No
>
> If yes, why?
> _____

Figure 4: *Information card T*

card	group 1	group 2	group 3	group 4	group 5	group 6	group 7	group 8	group 9
2.01									
2.02				S					
2.03	J	J	J	J	J	J	J	J	J
2.04									
2.05									
2.06									
2.07	M, K	M	M, K	M	M	M	M, K	M, K	M, K
2.08				G	K				
2.09	H3	H3	H3	H3			H3	H3	H3
2.10									
2.11	T		T	T	T	T	T	T	T
2.12	V	V	V	V	V	V	V	V	V
2.13	W	W	W	W	W	W	W	W	W, H4
2.14									

Table 1: *Distribution of cards and questionnaires during the second session*

The influence of students' individual experiences

The teacher also gave the students four questionnaires (see figure 1) in every laboratory session concerning specific elements of the learning environment. These were the teacher's only actions; she did not help, did not answer (physics) questions and did not give any hints.

For an overview of the cards and questionnaires distributed in the second session to nine groups (each of three students), see table 1. One group did not get the questionnaires and is therefore missed out.

The first column contains all cards numbered from 2.01-2.14. Every capital letter refers to an additional card (intervention, thought experiment or information) which was given to the groups while working on a numbered card. The shaded areas refer to questionnaires which were given to the students when they had just finished working on the relevant card(s).

Results of the questionnaire

Figure 5 presents the distribution of mean values of students' answers concerning 15 elements of the learning environment, sampled from all three sessions. These elements are the numbered cards, or numbered cards plus additional cards, in all cases where three or more groups completed the

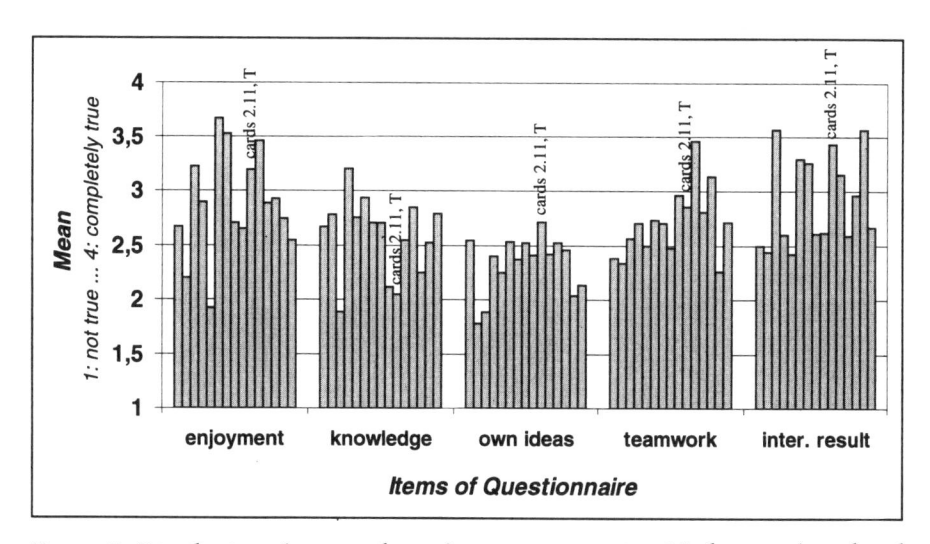

Figure 5: *Distribution of mean values of answers concerning 15 elements (numbered cards or numbered cards plus additional cards) of the learning environment, sampled from all three sessions*

questionnaire (for examples see table 1). Every tenth column for each item refers to questionnaires concerning cards 2.11 (figure 3) and T (figure 4), every eleventh column refers to cards 2.13, W, H4 and so on.

Figure 3 shows that the students enjoyed most of the elements (cards connected with tasks and experiments) and normally considered the result to be interesting. The students often thought that they knew all that was needed to succeed in the experiment (task). Working together with a part-

name	enjoyment	knowledge	own ideas	teamwork	int. res.
Alexa	2	2	2	3	3
Natalie	4	2	2	3	4
Tina	4	2	3	2	4
Mario	4	3	3	2	4
Christoph	2	4	4	2	2
Jochen	4	1	2	3	3
Katharina	3	1	2	2	4
Kerstin	2	1	1	3	3
Sonja	3	2	2	3	4
Tim	2	3	3	3	3
Marc	3	2	1	2	1
Eva-Maria	3	1	4	3	4
Henriette	4	1	4	1	4
Allin	4	1	1	3	4
Johanna	4	1	1	3	4
Kim	3	2	2	3	3
Stefan	2	4	4	4	2
Anne	3	2	4	4	4
Tom	4	3	4	4	4
Jonas	3	2	4	4	4
Malte	4	3	4	3	4

Table 2: *Distribution of answers concerning card 2.11 and card T (1: not true, 2: partly true, 3: mostly true, 4: completely true)*

The influence of students' individual experiences

ner student was not thought so important. The students often thought, that they could not introduce their own ideas into the experiments (tasks).

To explore in more detail the students' experiences with the experiments, this analysis of the distribution of mean values of answers is not very helpful.

Table 2 shows the distribution of answers concerning a specific element of the learning environment: cards 2.11 and T (see figures 3, 4). The students were asked to find out which end of the neon bulb indicates which charge. They were allowed to ask for further information (and almost all groups did so, see table 1). If they did, they received an explanation of how the neon bulb works (see figures 4, 6).

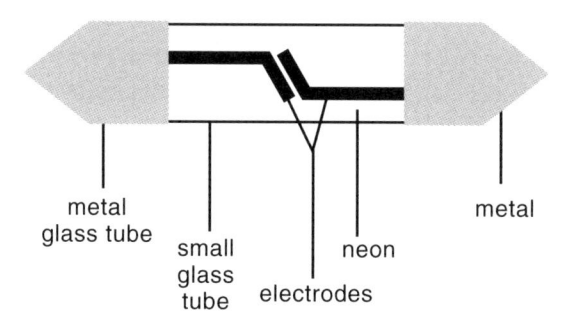

Figure 6: *Neon bulb*

After they had finished working on the cards they were given the questionnaire. The distribution of answers shows that there are groups who thought that they knew everything that was needed to solve the task successfully (e.g. Tom, Jonas and Malte, see table 2) and others who did not think so (e.g. Henriette, Allin and Johanna or Katharina, Kerstin and Sonja). Comparing the variation of answers between the groups we see similar results concerning the items 'own ideas' and 'teamwork'. There is also variation in the answers within the groups. For example Eva-Maria and Tim considered the result to be interesting, but Marc did not.

The distribution of answers of a single student (table 3) shows that there are items with (nearly) consistent answers ('teamwork') and items with variation in the answers ('knowledge' and 'own ideas'). For 'enjoyment' and 'in-

cards	enjoyment	knowledge	own ideas	teamwork	int. res.
1.02	4	2	3	1	3
1.11, A, B	4	3	2	1	4
1.20	3	4	1	1	3
1.21	4	4	1	2	4
2.07, M	3	4	3	1	3
2.09, H3	1	1	3	1	1
2.11, T	4	1	4	1	4
2.13, W	4	1	4	2	4
3.04	2	4	3	1	2
3.10	4	1	4	1	4
3.11	4	3	1	1	4
3.12, Y	1	1	1	1	1

Table 3: *Distribution of a single student's answers over the 12 questionnaires (1: not true, 2: partly true, 3: mostly true, 4: completely true)*

teresting result' most answers are similar but three elements (cards 2.09, H3, card 3.04 and cards 3.12, Y) are out of line.

Summarising the results described, we can say that the experience of a learning environment is highly context dependent and individual. Students' answers concerning different elements of the learning environment vary (see figure 5). We also find (differing) variation in the answers between and within groups concerning one element (see table 2), and we find different variation in the answers of single students concerning different elements of the learning environment (see table 3). We have not so far found any systematic variations. Without analysing the process data (students' (verbal) actions within the learning environment) we cannot say what might have caused these (unsystematic) variations.

Analysis of process data

Our group has a long tradition in analysing students' cognitive development by a detailed analysis of students' actions in physics environments, leading

The influence of students' individual experiences

to a reconstruction of situated cognitions (as 'ideas') of a single student which lie 'behind' his/her actions. To obtain a quantitative description of the development of these ideas we use levels of complexity (for a detailed description of the method and the levels see von Aufschnaiter, S. & Welzel 1997a/b; von Aufschnaiter, C. 1999; Welzel, von Aufschnaiter, C. & Schoster, and Schoster & von Aufschnaiter, S. in this section). We have found out that

- situated cognition of learners in new learning environments always starts with manipulating objects.
- situated cognition develops from manipulating these objects to the abstract description of properties of classes of objects and to the identification of connections between these properties.

I want to know which end of the neon bulb refers to which charge.	The neon bulb flashes at the end which touches the negative part of the object.	If you rub two objects together one will be negatively charged and one positively.	If the side of the neon bulb that is away from the object flashes, the object has a positive charge.	Flashing on different sides refers to different kinds of charges.
The charge of an object depends on the object's material.	A neon bulb flashes when it is in contact with a charged object.	The neon bulb can flash at different ends.	A neon bulb can distinguish between positive and negative charge.	Using a neon bulb you can distinguish whether an object is positively or negatively charged.
I rub different materials to see which end of the neon bulb flashes when I touch the rubbed objects.	The neon bulb flashed at the back.	I never saw the neon bulb flashing at the front side.	Only one side flashed.	I have to find out which end of the neon bulb flashes.

Figure 7: *Examples of students' situated cognitions concerning the neon bulb and development of these situated cognitions with respect to frames*

For the tasks presented on the neon bulb (see figures 3, 4, 6), examples of the students' (reconstructed) situated cognitions are shown in figure 7 (for a detailed description of students' situated cognitions about the neon bulb, see Schoster & von Aufschnaiter, S. in this section). We ascribed, for example, to a student's activity the idea: 'The neon bulb can flash at different ends' or the idea 'If you rub two objects together one will be negatively charged and one positively' (see figure 7). The examples presented in figure 7 are at three different levels of complexity (each row refers to one level).

If we describe how students develop situated cognitions, we see that students frame tasks. Frames are students' situated cognitions concerning the construction of a rough image of what students want to find out or where the task will lead to. Therefore frames are constructed on a three second time scale. These frames orientate students' following actions on time scales of sequences of Images-of-now (a few minutes). Every constructed frame refers to contents encountered previously and is at a level of complexity which has been achieved in prior situations, and to an estimation of what can be attained in the present situation. From an observer's point of view frames can be reconstructed as ideas on the basis of students' actions and utterances.

Figure 7 shows two frames that refer to card T (see figure 4). With respect to the task one student wants to find out which end of the neon bulb indicates which charge (see figure 7, light grey rectangle). The other student wants to find out which end of the neon bulb lights up (see figure 7, dark grey rectangle). These are two different frames. Students develop situated cognitions with respect to their frame. For the frames presented above, one student developed situated cognitions at one complexity level (see figure 7, dark grey arrows), whilst the other developed situated cognitions to a higher complexity level than he had started with (see figure 7, light grey arrows).

As we cannot present all the situated cognitions these students constructed while acting with the neon bulb, these examples are introduced only to show the 'general' idea of development. A lot of situated cognitions have taken place in between these examples. If students attain their frame by developing situated cognitions, they construct a new frame (e.g. referring to the next task, or some private things). That means that in every situation, action is always connected to frames. Students develop situated cognitions at one complexity level, or build from the 'bottom-up' to higher complexity levels with respect to their frame (see also von Aufschnaiter, S. & Welzel 1997b). By observing students we have found that the attainment of a frame occurs within periods of about five minutes. Within this period students finish the task by reaching their frame successfully (as shown in both examples in figure 7), or they give up because the think that they cannot reach their frame.

Interdependencies between individual experiences and cognitive development

In the previous sections we have shown that answers to the questionnaire show strong variations as regards the students' individual and context related experiences which may be explained by analysing the process data. Analysis of process data leads to the description of a single student's development of situated cognitions. Comparing this development and the students' situated actions and utterances (about enjoyment, failure, success, interest etc.) with their answers to context related questionnaires leads to initial results and hypotheses concerning the interrelation of cognitive development and individual experiences. A feeling of enjoyment depends on acting successfully with respect to the frame and developing situated cognitions at the same complexity level. This means that a feeling of enjoyment is not a consequence of discovering 'new' things but of developing and improving 'routines' in 'familiar' contexts.

The students consider results to be interesting when they reach their frame and can develop situated cognitions to higher complexity levels than they had started with. In these situations the students often indicate their own perception of their 'learning progress' in the last two questions of the questionnaire (see figure 1) and during the tasks. They say things like 'I have found out something new', 'I learnt something new', 'I was able to answer my question' or 'I got new knowledge'. Taking these comments and the situated cognitions to which they refer into account leads us to a hypothesis about feeling of competence. Students' feeling of competence refer to discovering new things and meanings. That means that interest in tasks has a close relation to the development of feeling of competence.

The analysis of the students' opinions about the importance of teamwork shows that students often consider interactions to be necessary for developing competence but not always for enjoyment. Our hypothesis is that students' opinions about the importance of teamwork depend more on the results they achieve than on the intensity of their teamwork.

A summary of our results and hypotheses shows that developing 'routines' is enjoyable and that getting to know new things and constructing new meanings is interesting and gives a feeling of competence (von Aufschnaiter, C. 1999). In both cases the frames play an important role. From students' point of view being successful is not decided within a few situated cognitions on time scales of seconds. Success is experienced if sequences of Images-of-now lead to reaching the frame within a few minutes. This time scale may be relevant for learning processes (changes in the cognitive system).

Taking time dependent dynamics into account the most important hypothesis is that students will have no enjoyment, will not feel competent and will not consider anything to be interesting when they cannot reach their frame within around five minutes. The same happens when the frame is reached within a few seconds: then students think that the task was boring.

Consequences for learning environments

(Normally) teachers like to have happy students who feel competent, act with enjoyment and consider the learning environment to be interesting. However, it is 'hard work' to construct learning environments which support these positive experiences. A learning environment should comprise alternatives which allow different students to construct frames of different complexity levels. As students' frames always refer to contents and are at levels of complexity which have been achieved in prior situations, the learning environment has to be adapted to students' momentary and individual development of situated cognitions in order for framing to be possible. Tasks have to be subdivided (or to be subdividable by students) so that (parts of) the task can be solved successfully by students within a period of around five minutes. While working on (parts of) the task students should have the chance to act autonomously.

In recent years many studies have been carried out to explore the decrease of students' interest in natural science subjects, especially physics and chemistry (e.g. Häussler 1987; Baumert et al. 1997). We think that a fundamental requirement for the development of a permanent interest in a subject is that positive experiences should predominate. Therefore we have to support students to find enjoyment, to experience competence and to consider tasks to be interesting.

References

Baumert, J. et al. (1997): *TIMSS – Mathematisch-naturwissenschaftlicher Unterricht im internationalen Vergleich: Deskriptive Befunde.* [TIMSS – Mathematical and Natural Science Teaching: International Comparision] Opladen: Leske + Budrich

Brown, J. S., Collins, A. & Duguid, P. (1989). Situated Cognition and the Culture of Learning. *Educational Researcher,* 18, 32-42

Clancey, W. J. (1993). Situated Action: A Neuropsychological Interpretation. Response to Vera and Simon. *Cognitive Science,* 17, 87-116

Csikszentmihalyi, M. & Larson, R. (1987). Validity and Reliability of the Experience-Sampling Method. *The Journal of Nervous and Mental Disease,* 175(9), 526-536

Damasio, A. R. (1994): *Descartes' Error. Emotion, Reason, and the Human Brain.* New York: Avon Books

Deci, E. L. (1992). The Relation of Interest to the Motivation of Behavior: A Self-Determination Theory Perspective. In Renninger, K. A., Hidi, S. & Krapp, A. (eds), *The Role of Interest in Learning and Development.* Hillsdale, N. J.: Lawrence Erlbaum, pp. 43-70

Deci, E. L. (1995): *Why We Do What We Do. Understanding Self-Motivation.* Harmondsworth, Middlesex, England: Penguin

Deci, E. L. & Ryan, R. M. (1993). Die Selbstbestimmungstheorie der Motivation und ihre Bedeutung für die Pädagogik. [The Self-Determination Theory and its Implication for Educational Science.] *Zeitschrift für Pädagogik,* 39(2), 223-238

Fischer, H. E. & von Aufschnaiter, S. (1993). The Development of Meaning During Physics Instruction. *Science Education,* 77(2), 153-168

Häussler, P. (1987). Measuring Students' Interest in Physics – Design and Results of a Cross-Sectional Study in the Federal Republic of Germany. *International Journal of Science Education,* 9(1), 79-92

Krapp, A. (1996). Die Bedeutung von Interesse und intrinsischer Motivation für den Erfolg und die Steuerung schulischen Lernens. [The Importance of Interest and Intrinsic Motivation for Success and Control of Educational Learning.] In Schnaitmann, G. W. (ed), *Theorie und Praxis der Unterrichtsforschung.* Donauwörth: Auer, pp. 87-110

Lave, J. (1997). The Culture of Acquisition and the Practice of Understanding. In Kirshner, D. & Whitson, J. A. (eds), *Situated Cognition: Social, Semiotic, and Psychological Perspectives.* Mahwah, N. J.: Lawrence, pp. 17-36

Pöppel, E. (1994). Temporal mechanisms in Perception. *International Review of Neurobiology,* 37, 185-202

Renninger, K. A., Hidi, S. & Krapp, A. (eds) (1992): *The Role of Interest in Learning and Development.* Hillsdale, N. J.: Lawrence Erlbaum

Schiefele, U. & Schreyer, I. (1994). Intrinsische Lernmotivation und Lernen. Ein Überblick zu Ergebnissen der Forschung. [Intrinsic Learning Motivation and Learning. An Overview about Results of Research.] *Zeitschrift für Pädagogische Psychologie*, 8(1), 1-13

Schoster, A. & von Aufschnaiter, S. (1998). Der Einfluß unterschiedlich komplexer Lernumgebungen auf die Lernentwicklung. [The Influence of Different Complicated Learning Environments on Learning Processes.] In *Zur Didaktik der Physik und Chemie: Probleme und Perspektiven*. Hrsg. von der GDCP, Kiel. Alsbach/Bergstraße: Leuchtturm-Verlag, pp. 343-345

von Aufschnaiter, C. (1999): *Bedeutungsentwicklungen, Interaktionen und situatives Erleben beim Bearbeiten physikalischer Aufgaben. [Situted Cognitions, Interactions and Situated Experiences While Working on Physics Tasks.]* Dissertation am Fachbereich I (Physik/Elektrotechnik) der Universität Bremen. Berlin: Logos

von Aufschnaiter, S. & Welzel, M. (1996). Beschreibung von Lernprozessen. [Description of Learning Processes.] In Duit, R. & Rhöneck, C. v. (eds), *Lernen in den Naturwissenschaften*. Kiel: Institut für die Pädagogik der Naturwissenschaften, pp. 301-327

von Aufschnaiter, S. & Welzel, M. (1997a). Wissensvermittlung durch Wissensentwicklung. Das Bremer Komplexitätsmodell zur quantitativen Beschreibung von Bedeutungsentwicklung und Lernen. [Instruction through Knowledge Development. The Bremen Complexity Model for a Quantitative Description of Development of Situated Cognition and Learning.] *Zeitschrift für Didaktik der Naturwissenschaften*, 3(2), 43-58

von Aufschnaiter, S. & Welzel, M. (1997b). Individual Learning Processes – a Research Program with Focus on the Complexity of Situated Cognition. *Proceedings of the 1st European Conference of ESERA,* September 2-6, 1997, Rome -in press

Welzel, M. (1997a). MACROBUTTON HtmlResAnchor Investigations of Individual Learning Processes – a Research Program with its Theoretical Framework and Research Design. In *Proceedings of the 3rd European Summerschool*. Theory and Methodology of Research in Science Education, pp. 76-84

Welzel, M. (1997b). Student-Centred Instruction and Learning Processes in Physics. *Research in Science Education,* 27(3), 383-394

Wild, K.-P., Schiefele, U. & Krapp, A. (1995). Course-Specific Interest and Extrinsic Motivations as Predictors of Specific Learning Strategies and Course Grades. *Paper presented at the 6th European Conference for Research on Learning and Instruction*, Nijmegen

The influence of learning environments on cognitive processes

Anja Schoster and Stefan von Aufschnaiter

Our project focuses on the learning processes of students being instructed in physics. We have been examining, in lab situations, the effect of different complicated learning environments on the dynamic of individual learning processes. The analysis shows the variations of cognitive developments of different students to understand physics explanations. With our results we demonstrate that a learner has to start at the object level when solving a new task or understanding sentences in a physics context, and needs a lot of time to construct cognitions on increasingly higher levels of complexity All situated cognitions have a certain complexity. So the learner needs time to construct other/ 'new' situated cognitions (in the same content area) on higher complexity levels. By contrast, the same sentences or tasks in a physics context may appear to be easy for an expert like a physics teacher because he/she is already able to construct situated cognitions on a high complexity level just in time (within a few seconds). With these results in mind a good learning environment has to consist of a sequence of tasks beginning with low complexity and then to increase step by step to higher complexity.

Introduction

The predominant hypothesis found in the literature to date is as follows: in a specific content area like electrostatics students are guided by a few, fundamental, and rather complex conceptions or concepts when solving physics problems (diSessa and Sherin 1998). The change from misconceptions to 'right' physics concepts is considered as the main aim of physics instruction or practical work. Specific teaching strategies for the initiation of these changes are described at a basic level (see Scott, Asoko & Driver 1991). Today, in the theory of conceptual change, the elimination of misconceptions is not the only focus of the theory (Duit, 1996). It should be obvious for students that in specific situations preconceptions do not offer adequate orientations, and in these situations science conceptions have to be used (Duit 1996). Duit (1996) describes conceptual change as follows: 'Learning can be seen as a process of the cognitive development which derives from specific pre-school conceptions and changes to science conceptions. This process is already contacting existing cognitive structures' (p. 146). Duit (1996, 150; translated by the authors) and other authors mention four conditions required in a learning situation in order to initiate a conceptual change (see Beeth 1998; Posner et al. 1982): dissatisfaction on the side of the learners with the 'old' conception, and the existence of a new conception which seems to be intelligible, plausible and fruitful for the learners. Hewson & Hewson (1992), however, describe conceptual change as a change of status in conceptions towards more scientific conceptions. Thornton (1994) has developed a phenomenological framework and a methodology in order to identify student views of the physical world and to explore the dynamic process by which these views are transformed during instruction. This approach, 'Conceptual Dynamics', provides a method for the ordering of student views into learning hierarchies. He hypothesized a model for the transition from one view to another that postulates that many students move through a transitional state when changing views. diSessa and Sherin (1998) introduce coordination classes as a particular type of concept which is a systematic collection of strategies for reading a certain type of information out from the world.

Amongst a range of ongoing studies, we have been investigating the way of thinking of different students solving the same task, and the same students solving different tasks, in laboratory studies. As one result of these investigations we cannot confirm the hypothesis that students only activate a few complex conceptions while solving a problem. So, our result is that students normally generate a lot of different situated cognitions in an ad hoc way. 'Situated', as used here, primarily means situated in time, i.e. every 3 seconds a new situated cognition is produced by the cognitive system connected to a greater or lesser extent among other things with changes in the learning environment (von Aufschnaiter, Schoster & von Aufschnaiter, in this section).

We have developed a theoretical frame and specific analysing methods for describing the variety of different ways of student thinking and for identifying the dynamic of situated cognition development so far.

Theoretical frame and research questions

For more than 10 years we have been investigating normal physics instruction which focused on the development of situated cognition amongst individual students. In this context we found that the complexity level of situated cognition of a student and the dynamic of the development of his/her 'new knowledge' depends on the *complicatedness* of the physics instruction, on the interaction with the teacher and fellow students and on the individual experiences of the student being observed. For the investigation of cognitive development it was essential to describe the cognitive level of situated cognition (von Aufschnaiter & Welzel, 1996). Each individual situated cognition can be seen in a close interrelation to the learning environment. Accordingly, individual situated cognition always refers to specific contexts, resulting in more appropriate actions. So the learning environment, in particular the complicatedness of a specific task, has to be classified. We distinguish between *complexity* of situated cognitions (what is instantaneously going on in a student's mind while working on a task) and the *complicatedness* of the task (that means what a complexity level every student has to reach to solve the task correctly). In our approach the term complicatedness is therefore used to describe the learning environment (the tasks) whereas the term complexity describes the learner's situated cognitions (that refer to the learning environment). Normally a student when trying to solve a task starts situated cognitions on a lower level of complexity as is necessary, and tries within sequences of situated cognitions to reach the level of complexity being necessary for solving this task.

Thus, a major question of our investigation is related to the development of *the complexity of cognitive processes* in a more or less complicated *learning environment*.

The following model of complexity levels (figure 1) shows the main analysing instrument of our investigations (von Aufschnaiter & Welzel 1997).

The complexity level of single situated cognitions can be empirically described using this heuristic method.

Using this design, which is described in more detail by von Aufschnaiter & Welzel (1997), we found a lot of examples in a number of different studies which show that situated cognitions always develop bottom-up from lower to higher complexity. Furthermore we observed that students only accept offers given by their teacher or fellow students for further actions which cor-

Levels of complexity	Description
Systems	Construction of stable networks of variable principles
Networks	Systematic variation of a principle according to other principles
Connections	Links between several principles with the same or different variable properties
Principles	Construction of stable covariations of pairs of properties
Programmes	Systematic variation of a property according to other properties
Events	Links between some stable properties of the same or of different class(es) of objects
Properties	Construction of classes of objects on the basis of common and different aspects
Operations	Systematic variation of objects according to their aspects
Aspects	Links between objects and/or identification of specific features
Objects	Construction of stable figure-ground-distinctions

Figure 1

respond to the actual complexity level of their situated cognitions or which are lower in complexity (Schoster & von Aufschnaiter 1998). These results come from case studies in regular physics courses, taught up to 1996, where it was not possible to plan series of tasks with respect to their complicatedness. In a preliminary laboratory study we analysed processes in which the learning environment is well defined in this relation. Ten groups (three students each) were presented with the same series of more than 50 problems of electrostatic phenomena in three double lessons. For the predetermination of the complexity of each task, we raised the complexity level on which students had finally solved this task in a similar learning situation. We attributed the level of complicatedness to the task on the basis of this complexity level. It then becomes possible to study differences and common features of the dynamic of situated cognitions with respect to the development of complexity of situated cognitions of different students when trying to solve this task.

As an example from this study, we now consider the question 'What cognitions must be necessarily developed situatively by the students to understand the flashes of the neon bulb?' in detail.

General conclusions given at the end of this paper are based on the cognitive development of all 30 students and 51 tasks from this study.

The design of the investigation

In normal physics instruction, the complicatedness of the learning environment and instruction is not planned with any regard to the complexity level of the situated cognition of each student. We set great store by this in the presented investigation. The complicatedness of the physics instruction was determined in advance on the basis of previous empirical investigation (Schoster 1998).

As our learning environment, we planned three double lessons in which students (age 17) do electrostatics in groups (three students per group, altogether ten groups).

The lessons were structured by tasks with increasing complicatedness, written on cards. During the lessons the students had to solve 51 tasks in which they were asked to do experiments and to explain the results they had observed (for an example see figure 2).

– Tasks –
➥ to use in the given order
➥ increasing complexity
➥ containing labwork instructions and questions on electrostatic phenomena

Example:

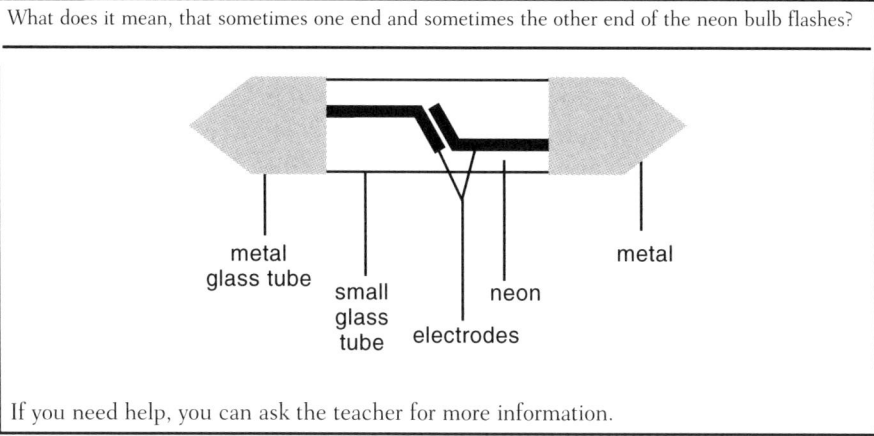

What does it mean, that sometimes one end and sometimes the other end of the neon bulb flashes?

metal glass tube   small glass tube   neon   electrodes   metal

If you need help, you can ask the teacher for more information.

Figure 2

A room was prepared for the students at the university (figure 3). Here they sat together around a table. A box with material for electrostatic experiments was given to them.

Different materials were prepared for the implementation of electrostatic experiments including transparencies, dusters, little balls made of polystyrene, threads, tennis balls with and without a graphite coat, small pieces of paper, neon bulbs, electroscopes etc.

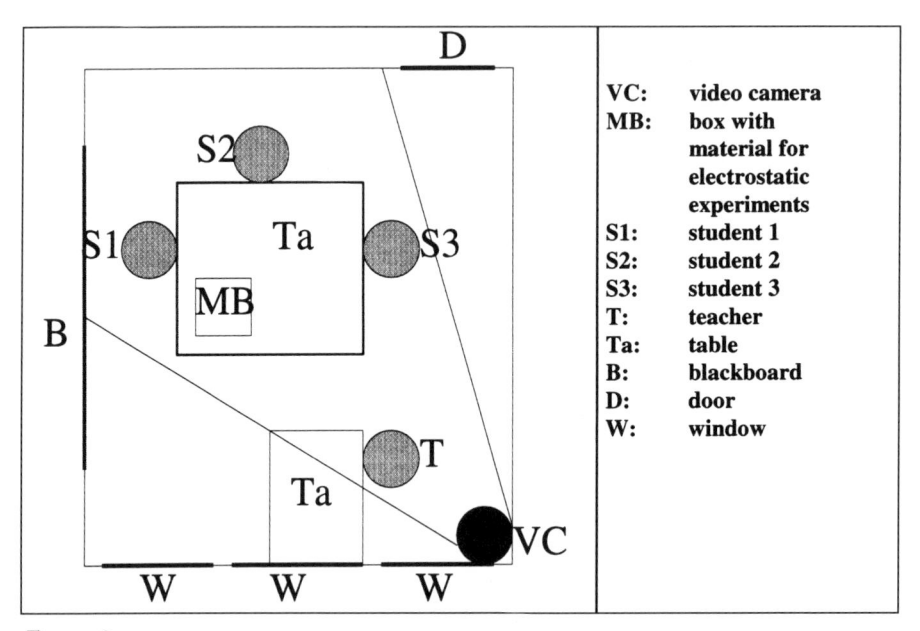

Figure 3

The structure of the instruction material is shown in figure 4.

During their experimental activities the students were given additional cards. In some cases 'information cards' were offered to students, which they could ask for (for an example see figure 5).

Moreover 'hypothetical experiments' (written on cards too) were given by the teacher to students who could not see the phenomena they should have observed, and 'intervention cards' of much higher complicatedness were distributed by the teacher at a fixed schedule for being able to analyse how students use these suggestions. Here we describe only how one of the information cards influenced the cognitive processes of the students when solving one task.

The influence of learning environments on cognitive processes

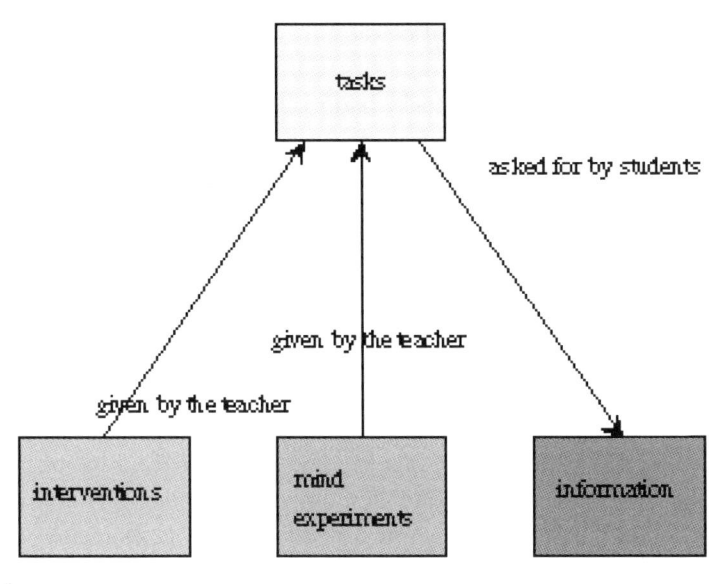

Figure 4

– Information –
➥ asked for by students
➥ low complexity
➥ containing explanations of the physics terminology and of the function of equipment

Example:

Explanation of the neon bulb

Using a neon bulb you can differentiate whether an object is positively or negatively charged. The neon bulb flashes at the end which touches the negative part of the object.

Do you think, that this information can help you by solving future tasks?

☐ Yes   ☐ No

If yes, why?

Figure 5

On average, seven additional cards were handed out by the teacher per double lesson. These were the only situations in which the teacher communicated with the students. All tasks had to be solved by the students themselves without getting any help from the teacher.

Data and data analysis

The students were continuously videotaped during their labwork. To analyse the videodata we produced transcripts. All statements and gestures of the students (and the teacher) are written down in chronological order. With these transcripts we reconstruct what situated cognition the student may have developed to carry out the observed actions. For these reconstructions we use the term 'ideas'. Each idea embodies normally one (sometimes two or three very similar) situated cognitions (3 seconds Images-of-now, von Aufschnaiter, Schoster & von Aufschnaiter, in this section) of the observed student.

These ideas are analysed with respect to their complexity (von Aufschnaiter & Welzel 1996; von Aufschnaiter & Welzel 1997; Welzel 1997, 1998).

Using an example we want to explain the way we set up a list of ideas using the transcripts and how we interpret these lists of ideas.

For that purpose we show a part of a transcript (figure 6) which includes the students Henriette, Allin and Johanna. The working on the cards (figure 2 and figure 5) needs 7 minutes. Henriettes' ideas will be discussed in detail.

The situation

In the presented situation Johanna reads the task (figure 2) aloud. Then Henriette remembers, from a previous card, that it is always the end of the neon bulb which touches the charged object that flashes (figure 6, line 1.2-1.3), which is in contradiction to the instruction on the task card (figure 2). This task card refers to an information card (figure 5) and the students ask for it. Henriette reads it aloud (figure 6, line 2.1-2.4).

Henriette tries to interpret the different parts of this 'information' together with the other two girls, but she cannot understand the meaning. Oriented by a question of Johanna, they then discuss the meaning of a negative or a positive charge with regard to a lack or a surplus of electrons. But they

1.1	Henriette:
1.2	They show us that it is always the end of the neon bulb flashes which isn't
1.3	touching the charged object.
	(...)
2.1	(reading) Explanation of the neon bulb: You can determine with a neon bulb
2.2	whether an object is positively or negatively charged (stops reading). Really?
2.3	(very astonished). (reading) **The neon bulb always flashes at the end which**
2.4	**touches the negatively charged object** (stops reading). I didn't know that.
2.5	(looks at the card, repeats indistinctly, reads again) ...touches the negatively
2.6	charged object. Negatively charged ... negatively ...That's right!
2.7	Allin:
2.8	There is a surplus of electrons or a lack? I don't know.
	(...)
3.1	Oh, then **we have done all the experiments wrong**. (looks at the information
3.2	card) Wait! Yes, because we don't know this.
3.3	Johanna:
3.4	But the other end of the neon bulb didn't flash.
3.5	Henriette:
3.6	Maybe all the objects were negatively charged. Or we carried out the experi-
3.7	ment wrong.
3.8	OK. Let's try it again, shall we? (she picks the metal plate with the handle as
3.9	well as the PVC-transparency) So when I rub these together, for example, one
3.10	must have a negative charge and the other a positive charge.
3.11	(She rubs metal plate and PVC- transparency together) We have to see now
3.12	which end of the neon bulb flashes. (Henriette gives the metal plate to Allin,
3.13	because she can not see the flashes of the neon bulb, they repeat the experi-
3.14	ment.)...Well, this is how I understand it. The object ...
3.15	Allin:
3.16	So when the object has a negative charge, the end of the neon bulb which
3.17	touches the negatively charged object flashes.
3.18	Johanna:
3.19	... away from the charged object ...
3.20	Henriette:
3.21	... at the side of the charged object ...
3.22	Allin:
3.23	... at the end touching the charged object
3.24	Henriette:
3.25	Yes, so you can conclude that if the other end flashes the object is positively
3.26	charged (pointing to the appropriate end of the neon bulb).
3.27	**Henriette (together with Allin):**
3.28	...then the object has a positive charge.

Figure 6

cannot connect these approaches of the different students with the flashes of the neon bulb.

After two minutes, Allin reads the task again. Henriette reads the information card and repeats the part that says 'negative charge'.

After that, Henriette and her group were able to answer the question posed on the task card and they finished their work on the two cards (figure 2 and 5).

Henriette's ideas in the transcript (figure 6)

Henriette's first idea (this refers to figure 6, line 1.2-1.3), 'The end of the neon bulb which touches the charged object always flashes', describes a property of a neon bulb. This idea does not contain a solution to the task but initiates other situated cognitions because Henriette sees a contradiction to the text previously read by Johanna.

In the next part of the transcript the first idea of Henriette is 'You can determine with a neon bulb whether an object is positively or negatively charged' (figure 6, line 2.1-2.2). In this idea she accepts what is written in the sentence as a possible property of a neon bulb. Henriette cannot understand the next sentence (figure 6, line 2.3-2.4). She sometimes repeats certain parts of this sentence. Allin initiates a new aspect for a discussion. After some time (two minutes) where the three students have discussed this aspect, Henriette concludes (figure 6, line 3.1): 'We have done all the experiments wrong'. Here she thinks about specific operations with certain materials. Her next idea refers to the negative charge of objects: 'Maybe all the objects were negatively charged' (figure 6, line 3.6). This idea describes a (possible) property of the investigated objects. But Henriette gives another (possible) explanation for the result of their experiments (figure 6, line 17-18): 'We carried out the experiments wrong'. This idea also describes a property related to the implementation of the experiments.

The following three ideas refer to thought operations with specific objects (figure 6, line 3.8-3.10):

> 'We can make an experiment, rubbing two specific objects like a metal plate and a transparency together'
> 'If you rub two objects together one will be negatively charged and one positively'
> 'If you touch the two objects (one after another) then both ends of the neon bulb should flash'

The next ideas (figure 6, line 3.11-3.12) contain the performance of the operations she thought about:
'I rub the metal plate with a transparency'
'I have to find out which end of the neon bulb flashes'
'The neon bulb flashes at the end which does not touch the charged object'

Henriette repeats the experiment once again and then she is able to construct a property that is new to her:

A neon bulb can flash at the end which does not touch the charged object (figure 6, line 3.13-3.21).

At the beginning of this sequence (figure 6), Henriette is able to construct the flashes of the neon bulb at the touching side as a property of neon bulbs because she had seen it several times when acting with negatively charged objects and the neon bulb. But without having seen the flashes of the neon bulb at the other end, she could not construct the property of the neon bulb (to flash at the side which does not touch the charged object) which is necessary to answer the question posed on the task card. It took 3 minutes to construct the 'new' property by operating again and again with the neon bulb and different materials.

Results

As demonstrated in the example of Henriette, we analysed the cognitive processes of all 30 students. We got about 250 different ideas altogether from these 30 students working on the task card (figure 2) and the information card (figure 5), 120 of them concerning the neon bulb.
 In addition to the example discussed for the student Henriette, we first describe her main ideas for the whole process of working on the cards (figure 2 and 5).
 As for Henriette some of these ideas could be 'seen' (by the observer) when the students were working on previous tasks with the neon bulb.

'A neon bulb flashes if it touches charged objects'
'A neon bulb flashes at the end that touches a charged object'
'There are objects with a positive and a negative charge'

'The neon bulb can flash at different ends'
'It has to do with a negative charge'

In the demonstrated sequence most of the students finally developed the new properties:

'Using a neon bulb you can distinguish between positive and negative charges'
'A neon bulb can flash at the end which doesn't touch the charged object'

Henriette and other students construct several connections of properties while working on the cards being necessary to answer questions on the task card, such as:

'A neon bulb can flash at the end which touches the negatively charged object'
'If the end of a neon bulb flashes that is away from the object, you can come to the conclusion that the object has a positive charge'
'If a neon bulb flashes at the end which doesn't touch a charged metal plate, then the metal plate has a positive charge'

We tried to make clear how we analysed the other 29 students solving the cards (figure 2 and 5) with the example of Henriette's development of situated cognition. Most of the 250 observed ideas of all 30 students were used by only a few students, but some of these ideas, which are very important to make progress in the development of situated cognition, were developed by nearly all of the students. We make the following points about these ideas:

1 *With regard to the use of specific objects:*
First, students explore objects and their features in real or imaginary operations. This corresponds to levels 1 to 3 in figure 1.

For the above example of the development of situated cognition, the following ideas are very important:

- The students have to see the flashes of the neon bulb at both ends when touching specific objects and they have to connect it with the feature of a specific object as being charged.

The influence of learning environments on cognitive processes

2 *With regard to properties of objects:*
Next, students construct an important feature as a property for a class of objects. This corresponds to level 4 in figure 1.

For the above example of the development of situated cognition, the following ideas are very important:

- A neon bulb can flash at different ends.
- The flashes of a neon bulb depend on the presence of charge.
- A neon bulb can be used to distinguish between a positive and a negative charge.

3. *With regard to links between properties:*
If a student is able to construct several properties of the same or different classes of objects she/he can link these properties. This corresponds to level 5 in figure 1.

For the above example of the development of situated cognition, the following ideas are very important:

- The neon bulb flashes at the end which touches a negatively charged object.
- If the end of the neon bulb that is away from the touched object flashes, this object has a positive charge.

So the 'stations' of the situated cognition development processes in our example correspond to the first five levels of complexity in figure 1. None of these students involved in this task reaches higher complexity levels in this situation.

Conclusion

The development of new knowledge in this paper first presented for the student Henriette and her group, and then for the other 27 students by example of a task (figure 2), is confirmed by all other tasks of this investigation and by normal physics instruction with other students in other case studies. The conclusions are drawn from all these cases.

- Students have to develop a lot of situated cognitions for a correct understanding of a (complicated) sentence or task.

- When trying to solve new tasks, students always have to construct situated cognitions first dealing with specific objects, notwithstanding whether these objects are real (in the learning environment) or remembered or invented (by the students). Then they may try to construct relevant properties of classes of these objects to connect two or more properties and to attribute special rules to the connected properties. Only when they are well trained on these levels in a specific context do they become able to construct more abstract cognitions like programmes (available properties) or principles (covariation of two variable properties) in this content area.
- Students need a learning environment which allows them to develop such situated cognitions 'bottom-up' in complexity with respect to a specific content. In this respect good learning environments have to consist of a sequence of tasks beginning with low complicatedness and then to increase step by step to higher complicatedness.
- Sentences or tasks in a physics context may appear to be easy for an expert like a physics teacher because she/he had learned to construct situated cognitions on a high complexity level without having to construct them on lower levels first. Nevertheless, when solving a new task every person has to start on the object level. He/she then needs time to construct cognitions on increasingly higher levels of complexity (we demonstrated this with the example of Henriette). Therefore a learner sometimes does not reach the complexity level necessary for a successful solution of the task, or may need several attempts at different times to find a successful solution.
- Students need a learning environment which offers a lot of different possibilities to construct many cognitions on low levels of complexity.

References

Beeth, M.E. (1998). Teaching for Conceptual Change: Using Status as a Metacognitive Tool. *Science Education*, **82**(3), 343-356

Damasio, A. R. (1994). Descartes' Error. Emotion, Reason, and the Human Brain. New York: Avon Books.

diSessa, A. A. & Sherin, B. L. (1998). What changes in conceptual change? *Journal of Science Education*, **20**(10), 1155-1192

Duit, R. (1996). Lernen als Konzeptwechsel im naturwissenschaftlichen Unterricht. [Learning as Conceptual Change in Science Education.] In R. Duit & C. v. Rhöneck (eds), *Lernen in den Naturwissenschaften*. Kiel: IPN, pp. 145-162

Hewson, P.W. & Hewson, M.G. (1992). The Status of Students' Conceptions. In R. Duit & F. Goldberg (eds), *Research in Physics Learning: Theoretical Issues and Empirical Studies*. Kiel: IPN, pp. 59-73

Posner, G.J., Strike, K.A., Hewson, P.W. & Gertzog, W.A. (1982). Accommodation of a Scientific Conception: Toward a Theory of Conceptual Change. *Science Education*, 66(2), 211-227

Schoster, A. (1998): *Bedeutungsentwicklungsprozesse beim Lösen algorithmischer Physikaufgaben. [Development of Meaning in Solving Algorithmic Physics Tasks.]* Dissertation am Fachbereich 1 (Physik/Elektrotechnik) der Universität Bremen. Berlin: Logos

Schoster, A. & von Aufschnaiter, S. (1996). The Influence of Different Complex Learning Environments on Individual Learning Processes. In: *Proceedings of the 3rd European Summerschool*. Theory and Methodology of Research in Science Education. Barcelona: Apunts BCN, 150-154.

Schoster, A. & von Aufschnaiter, S. (1998). Der Einfluß unterschiedlich komplexer Lernumgebungen auf die Lernentwicklung. [The Influence of Different Complicated Learning Environments on Learning Processes.] In: H. Behrendt (eds), *Zur Didaktik der Physik und Chemie, Probleme und Perspektiven*. Alsbach/Bergstrasse: Leuchtturm Verlag, pp. 343-345

Scott, P., Asoko, H.M. & Driver, R. (1991). Teaching for conceptual change: A review of strategies. In R. Duit, F. Goldberg & H. Niedderer (eds), *Research in Physics Learning: Theoretical Issues and Empirical Studies*. Kiel: Institut für die Pädagogik der Naturwissenschaften an der Universität, pp. 310-329

Thornton, R. (1994). Conceptual Dynamics: Changing Student Views of Force and Motion. In C. Tarsitani, C. Bernardini & M. Vincentini (eds), *Thinking Physics for Teaching*. Rome

von Aufschnaiter, S. & Welzel, M. (1996). Beschreibung von Lernprozessen. [Description of Learning Processes.] In R. Duit & C. v. Rhöneck (eds), *Lernen in den Naturwissenschaften. [Learning in the Natural Sciences.]* Kiel: IPN, pp. 301-327

von Aufschnaiter, S. & Welzel, M. (1997b). Individual Learning Processes – a Research Program with Focus on the Complexity of Situated Cognition. *Proceedings of the 1st European Conference of ESERA,* September 2-6, 1997, Rome -in press

Welzel, M. (1997). Student-Centred Instruction and Learning Processes in Physics. *Research in Science Education*, 27(3), 383-394

Welzel, M. (1998). The Emergence of Complex Cognition During a Unit on Static Electricity. *International Journal of Science Education*, 20(9), 1107-1118

How to interact with students?
The role of teachers in a learning situation.

Manuela Welzel, Claudia von Aufschnaiter, and Anja Schoster

Investigating individual learning processes in the field of learning physics through experiments, one focus of our research interest is the influence of interactions on situated cognition. The main questions we wanted to answer are: (1) How do interactions influence cognitive processes? and (2) How should teachers interact with students to initiate learning in the intended way? To answer these questions videotaped sequences of interactive experimental classroom situations were analysed in detail. The results show that specific factors – content and complexity – influence the cognitive processes of students through interaction with a teacher. With this paper we will present selected data from experimental teaching situations to describe the influence of interactions on situated cognition and to derive possibilities for teachers to interact successfully within a learning situation involving experiments.

Theoretical background

Which of us – teachers and researchers – responsible for the teaching of a science discipline and using experiments for teaching does not recognise this experience: The students were given an experimental device and an appropriate task to solve, but they are not able to work on this task. They ask what to do and why to do it. A teacher explanation follows, well formulated and presented in a friendly way, and well structured. The students say 'Yeah. It is clear now'. The students must have understood, the teacher hopes. Now the students have to experiment and to formulate the answer to the problem in their booklet. But, they do not give a correct and complete answer or explanation. In the following discussion the teacher has to conclude that some of the students did not understand in the intended way what they have been doing all the time. What was going wrong there? Why did the students not understand the teacher's explanations? These questions lead us to ask for rules of interaction between teachers and students.

'A person's knowing of a conceptual domain is a set of abilities to understand, reason, and participate in discourse... Any particular activities that a person engages in or learns to perform are embedded in a conceptual ecology that has been developed within a community of intellectual work.' (Greeno 1991, p. 176)

Taking the cited perspective, learning can be seen in general as cognitive development through an apprenticeship in the practices of a culture (Brown et al. 1989, Rogoff 1990, Roth 1995). Apprenticeship is characterised by the opportunities for learning that are structured by participation in authentic practices. In each community, including schools, there are experts and novices interacting in order to organise learning. Conceptualising learning and teaching as interactions in a community of knowers (novices and experts) and seeing learners as independent agents (Agre 1993, Welzel & Roth 1998), the teacher has to be seen as a partner for interaction and for structuring learning resources such as environments, materials, discussions and hints to allow students to develop individual knowledge, to negotiate and to share it.

Usually, in contrary to this assumption, at school the job of a teacher is understood as acting as a medium for transmission of knowledge. However, many empirical studies show evidence indicating that the metaphor of teachers as a medium for the transmission of knowledge is wrong (Pfundt & Duit et al. 1991). Knowledge is not transmissible. Moreover, the teacher can only initiate specific activities of students during interactions (Welzel 1995) which allow students to construct and develop their own knowledge. But, the question is, how can this happen in a planned way — in the way intended by the teacher?

Using this theoretical background, we have been investigating the dynamic of situated cognition and learning of differently experienced students aged 10 to 25 years in the field of physics in experimental classroom situations in our research group for more than ten years (von Aufschnaiter & Welzel 1996; Welzel 1997). For these purposes we have developed a theoretical frame epistemologically similar to that of Clancey (1993). It may roughly be outlined with the demand: 'Try to understand cognition without using any approach of symbolic representation' (von Aufschnaiter & Welzel in press). As described in the paper of von Aufschnaiter, Schoster and von

How to interact with students?

Aufschnaiter (at the beginning of this section) 'situated' in this context means *situated in time*. From a neurobiological point of view the cognitive dynamic must be described as sequences of 'Images-of-now' (Damasio 1994). Each Image-of-now has to be 'produced' by the cognitive system (neuronal structure) within three seconds (Pöppel 1994). For experiences of a much larger time than three seconds, the cognitive system will produce sequences of Images-of-now which are connected to a greater or lesser extent, among other things, with changes in the learning environment.

Learning is correlated with 'the success' of sequences of Images-of-now when solving problems and leads to changes in the cognitive system (neuronal structure).

The aim of the investigations reported here was to analyse and describe in detail the individual processes of situated cognition and learning within physics learning environments on the short time scale of sequences of Images-of-now and the influence of several aspects of the environment on individual cognitive processes on that time scale.

Methodologically this results in a detailed analysis of individual actions and interactions in authentic learning environments — and this always means in social environments — comparable to the microgenetic method of Siegler and Crowley (1991). Following the activities of individual students during instruction and interaction in detail makes changes become visible: the number and type of elements the students use and relationships they construct (between these elements) during acting in a specific context increases (Welzel 1995; 1997). This development concerns their actions and their cognitions and can be described in terms of *increasing complexity of situated cognition* in a bottom-up-direction. We have been developing a heuristic that allows us to describe this kind of development by means of ten levels of complexity:

1. objects—construction of stable figure-ground-distinctions;
2. aspects—links between objects and/or identification of specific features;
3. operations—systematic variation of objects according to their aspects;
4. properties—construction of classes of objects on the basis of common and different aspects;
5. events—links between some stable properties of the same or of different class(es) of objects;

6 programmes—systematic variation of a property according to other properties;
7 principles—construction of stable co-variations of pairs of properties; connections—links between several principles with the same or different variable properties;
9 networks—systematic variation of a principle according to other principles;
10 systems—construction of stable networks of variable principles.

Taking this heuristic to analyse reconstructed students' behaviour (subdivided on a time scale of three seconds; Images-of-now) it becomes possible to follow the dynamic of situated action along this behaviour and with this the development of situated cognition during interaction in a learning environment (Welzel 1997, Welzel & Roth 1998). Using it, we will see the moments of students' reaction on a teacher's hint or question and we can qualify the complexity of the students' reaction.

Thus, *interaction*, the *dynamic of cognitive development* and *complexity* are the theoretical and methodological bases for our investigations.

Research questions

On the basis of the theoretical background described above, we analysed the influence of teachers' interactions on students' individual cognitive processes in detail to find rules of interaction in a teaching-learning-situation. On the basis of these rules possibilities to improve communication and interaction in teaching situations will be derived and discussed.

The two main questions for this paper are:

(1) *How do interactions influence cognitive processes with regard to the time scale, its content, and complexity?*; and
(2) *How to interact with students to initiate learning up to higher complexity levels in the intended way, in certain subject areas?*

Methods

In order to answer the research questions videotaped sequences of students and teachers acting in physics lessons were analysed in great detail (for the procedure see Welzel 1995, 1997; Welzel & von Aufschnaiter 1997). Videodata of a physics unit (15 double lessons) on electrostatics involving German High School students (aged 15), and university students (aged 21 to 23) acting in labwork sessions with a tutor and of sequences of private lessons on mechanics (age of students was around 17), were analysed. For this purpose, video-sequences with interactive situations were transcribed, and the students' cognitions and teachers' hints and questions were analysed according to their influence on students' situated cognition, the content of communication and the complexity levels reached.

Results

Analysing all the data we found that interactions influence individual cognitions on a short time scale of around 5 seconds (1). The results also show specific factors – content (2) and complexity (3) – influencing the thinking pathways of students through interaction with a teacher.

These results will now be examplified using data from the physics unit on electrostatics (taught in a grade 10 course of a German High School) and one example found in the literature. The literature example is used deliberately to demonstrate that the findings reported here are not artefacts produced with our own databases. The examples used in the paper are used only to *illustrate* the results found in a large number of interactive sequences detected and analysed in each of the data sets.

We will start with a general rule – the time scale of interactional influences. This will be followed by the content and complexity results.

1. Interactions influence cognitive processes on a short-time-scale

Analysing our data we found out that interactions always influence cognitive processes on a short time scale of around 5 seconds.

See the following example

Situation: during the unit on electrostatics – four girls – Caren, Jessica, Inga and Birgit are doing an experiment. (See figure 1). The duration of the whole sequence is 25 seconds.

Between two charged metal plates a thin isolator is placed. One of the plates is connected with an electroscope. The upper plate ten will be moved upward.

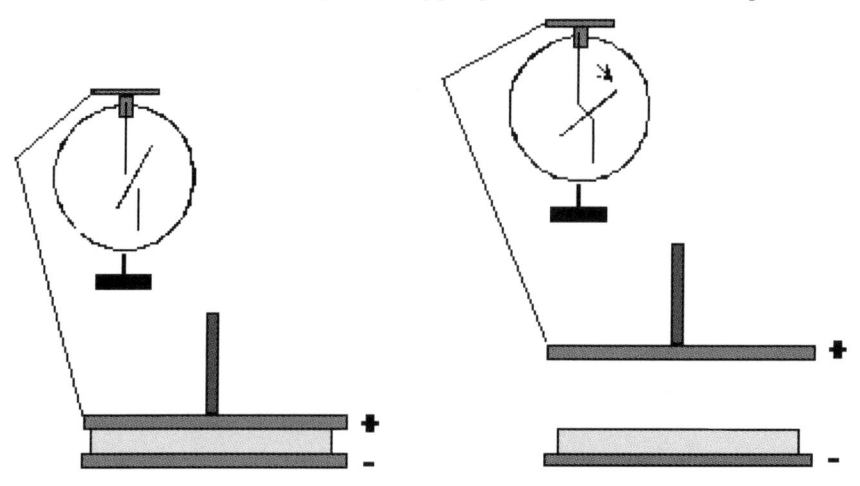

If the above plate is moved upward, the pointer at the electroscope moves forward.

Figure 1: *Experiment with metal plates and an electroscope*

The teacher approaches and asks the students:

Transcript (translated into English; T – teacher, C – Caren, J – Jessica, I – Inga, B – Birgit)

01 T: What happens, when enlarging the distance between the plates?
02 C: The pointer on the electroscope moves forward.
03 B: (MOVES THE UPPER PLATE) Yes, then the pointer moves
04 forward.
05 I: It enlarges, the pointer ...
06 T: Now you have to write the observation down, and do you know
07 why this happens?
08 I: Do you have the answering sheets? Or, ...
09 T: (POINTS AT A SHEET ON THE TABLE) Yes, write it down on this one
here.
10 I: O.K. Fine.
11 THE STUDENTS DO NOT WRITE.
12 T: Do you have an explanation for that?
13 I: That, because this is farther away now (MOVES HER
14 HAND AS THE UPPER PLATE), the electrons somehow

15		have to ...
16	B:	The electrons here don't push off themselves anymore
17		(POINTS TO THE UPPER PLATE)
18	J:	Yes, exactly.
19	B:	They are moving over there (POINTS TO THE
20		ELECTROSCOPE) and push off themselves.

What can be seen in this transcript?

In the first part (lines 02 to 05) the students answer directly to the question asked by the teacher. The teacher is satisfied by these answers and asks the students to write down the observations and to think about an explanation (lines 06-07). At this moment the student Inga asks for answering sheets. She focused on the first part of the teacher's question. The teacher answers this question, but in line 11 it can be seen that Inga does not react to this problem any more. She does not start to write anything. Because of no further action the teacher again asks for an explanation (line 12) to the pointer's movement. Immediately again Inga reacts: 'That, because this is farther away ...'.

A few seconds later (outside this piece of transcript) the students discuss the attraction of the electrons to the positively charged plate in relationship to the forces between the electrons on the surface of the upper plate. While Jessica is giving a complete and correct explanation relatively quick, the other three students need more time for a step-by-step-explanation. They do not use the explanation formulated by Jessica, although they attended the group all the time.

What about the time scale on which interactions influence cognitive processes? In our example (line 06 to 08) Inga did only react to the first part of the teachers intervention (where to write the observation down). The teacher had to formulate the second part of the intervention (explanation) again after the negotiation of the first part (line 12).

How do interactions influence the cognitive processes? Looking at the second part of the sequence the analyst can see that all explanations of the other students, especially the explanation formulated by Jessica did not produce a prompt right answer (explanation) of the other students of the group. Thus, each of the students needs his or her own way of constructing knowledge. And, the students need time to do so.

Manuela Welzel, Claudia von Aufschnaiter, and Anja Schoster

How to interact with students?

To answer this question we analysed all the data mentioned at the beginning. On the basis of many sequences and results, such as the example described above, we have to formulate (preliminary) general rules with regard to the *time scale* and the *dynamic* of the influence interactions can have:

- Ask only single questions, do not use complex sentences, *the students can only react to one aspect.*
- *Initiate only* further thinking, do not tell a long story, the students start to think about your comments (explanations) at the moment they can link parts to own experiences

Thus:

- Do not expect the students to remember all you or others said before, they *can remember what they 'experienced' by themselves only* or what they are able to use at the moment.

2. Content dependence

Our data show that interactions are successful only if the contents the students and the teacher 'talk about' (construct) are the same.

This statement seems to be trivial, but mostly, interactions of teachers during practical activities concern 'new aspects' of the experiment which the students cannot follow.

The following example taken from a chemistry lesson may illustrate this phenomenon (We took this example because it is from a different science discipline):

EXAMPLE:
The *'reduction strength'* of different metals is to be compared. The teacher puts an iron nail for a few seconds into a solution of copper (II) sulphate. He repeats the experiment using zinc (II) sulphate. Finally he puts a copper plate into a solution of silver nitrate.

How to interact with students?

Teacher: Let's observe whether we can see anything. What do you see?
Student: The nail is coated.
Teacher: The nail has become brown. In other words: obviously copper has been deposited there. Thus, copper ions are obviously able to react with iron.

What's happened with the nail in the zinc sulphate solution?

Students: (calling from around the room) It's smaller ... It seems to be brighter than the other ...
Teacher: Nothing has happened. Thus, obviously, the zinc ion is not able to react with iron. And the copper plate? ... You now observe the copper plate begins to take on a silver lustre. In other words, silver ions are able to react with copper. In this manner we could create a wonderful series for the metals.

(De Jong 1994, 180 – translated by the author)

Following this interaction between teacher and students one can see that both teacher and students communicate different things: The students describe more or less their observations with regard to the nails. In contrast, the teacher quickly communicates the chemical reactions. An observer cannot see whether or not the students really understand these explanations. With this it is not quite clear whether or not the students could follow the teachers' explanations and whether or not they are able to construct appropriate cognitions.

How to interact with students?

We found a lot of situations like the one described above in our own data sets where interacting partners have misunderstandings. Often the interacting partners do not recognise that a misunderstanding is taking place. To avoid interactions like this we propose that teachers follow these rules:

- *Start* your interaction *with the content where the students are* to change the direction together with them.
- If the students do not give the expected answer, *check if there are misunderstandings* based on different thinking contents.

3. Complexity

Interactions are successful only if student and teacher are at the *same level of complexity during interaction*. The levels of complexity we use for description are shown in figure 2 (and described in more detail in the Theoretical background of this paper):

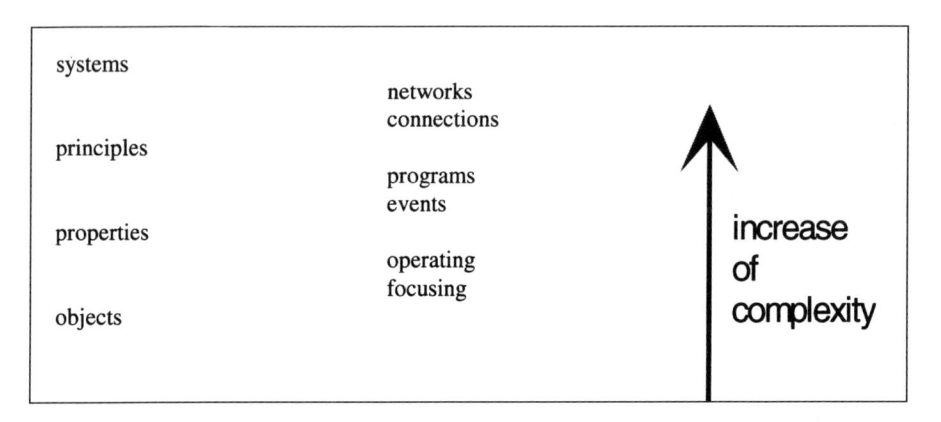

```
systems
                    networks
                    connections
principles
                    programs
                    events                    ↑
properties
                    operating               increase
                    focusing                of
objects                                     complexity
```

Figure 2: *Levels of Complexity*

Example for a successful Interaction

Let us go back to the first interactions of the example described above. In addition, now each statement has been analysed according to its complexity. The assigned level of complexity is noticed.

Example:
  The teacher approaches and asks the students, who are experimenting with the metal plates and the electroscope:

T: What happens, when *enlarging the distance* between the plates?
  (*She is varying the property of 'distance' – program*)
C: The *deflection* of the pointer at the electroscope *enlarges*.
  (*She is varying the property of 'pointer's deflection' – (program)*)
B: (MOVES THE UPPER PLATE) Yes, then the pointer moves forward.
  (*She is linking the lift of the plate to the pointer movement – event*)

I: It enlarges, the pointer ...
   (She is varying the objects 'plate and pointer' – operation)
T: Now you have to write the observation down, and do you know why ...
   (She is reacting to C's answer mainly)

The teacher at the beginning is formulating her question on a program level (constructing distance as a variable property). As the students were experimenting before and as they did vary the distance of the metal plate this level is appropriate to be used at this moment. It can be seen, that Caren (C) answers this question immediately at the same level (program). She talks about the change (variation) of the deflection. Both of them have been at the same complexity level.

But, looking at the answers of Birgit (B) and Inga (I) it seems to be that both of them did not reach the program level yet. They are at lower complexity levels, and the teacher does not notice it. She supposed that she had got the right answer from all of the three students. Thus, it seems to be obvious that the teacher goes further and asks for an explanation (see the full piece of transcript above).

This short sequence shows a partly successful interaction between teacher and the students.

Unsuccessful interactions

We have found a lot of similar situations with regard to unsuccessful interactions because of the mismatch of the complexity levels.

For example: In a teaching situation during the second lesson, after experimenting with different materials the teacher intends to repeat together with the students the experiences the students made during experimenting with different material. The students in the last lesson first made experiences at lower levels of complexity – up to the property level. The teacher starts to ask a 'triple-question' (see the transcript below) at a complexity level the students did not reach before – at the program level. Thus, the students are not able to understand what the teacher wants them to answer. The teacher tries to ask again, but at the same level. Until she asks at the property level, the students are unable to answer:

Manuela Welzel, Claudia von Aufschnaiter, and Anja Schoster

Transcript:
T: I ask you what happens while rubbing objects or, *how do these states of charge emerge?* Did someone of you hear about that before?
  (The teacher asks about variation of the property of charge – program)
No reaction
T: What happens *while charging?*
  (The teacher asks for linking the property of charge to other properties – event)
No reaction
T: Try *an explanation.*
  (This is the same as before – event)
No reaction
T: What happens *while rubbing?*
  (The teacher asks for linking concrete actions – operation)
Reaction   *(operation and property)*

How to interact with students?

Students pass a cognitive development when managing a situation. Within a series of other empirical studies we found out that this development is bottom-up in complexity and always starts at a low level (see Schoster & von Aufschnaiter in this section, and Lang 1998, Schoster 1998, Welzel 1995, 1997, Welzel & Roth 1997). Taking our example and, in addition, the results of the data analyses we mentioned, we summarise as follows:

- If you want to know what the students already know about a subject, *remember first what they did last time,* and *start to ask at a low level of complexity* (objects, properties...), guide them through the situation.
- If you want to give a hint during students' activities, *first observe* their activities and see *at what level they are*:
  - do they handle certain objects?
  - do they handle properties?
  - do they vary properties?
  Then formulate your hint *at the same level* (name or handle objects, name properties or focus on variation of properties).
- If you want to *guide students* to bring them towards a 'right way', look where they are, start there and then go further (in content and complexity). *The students cannot understand principles or laws at the beginning of*

a learning process.
- If tasks or problems need a too high level of complexity for solution, the students are not able to solve them. *Learners always interpret tasks on a complexity level they already reached before.*

Conclusions

Investigating individual learning processes in the field of learning physics with experiments on a constructivist basis one focus of our research interest was the influence of interactions in situated cognition on a short time scale. It was intended to extract concrete influencing factors (rules) and hints for teachers to behave appropriate within a learning environment involving experiments.

For this purpose, videotaped sequences of interactive experimental classroom situations were analysed in detail. The method of data analysis described was tested out for objectivity and reliability during the past 10 years with members of our institute.

On the one hand, our results show evidence that interaction exclusively on a short-time-scale influences situated cognition development of students. For interacting with students in a learning environment involving experiments, this leads to the recommendation to teachers only to ask single questions, and not to use complex sentences, because the students can react only to one aspect. Knowing this, a teacher exclusively can initiate a further thinking, therefore it is not useful to tell 'long and new stories', as the students always start to think about details of explanations at the moment they can link it to own experiences.

On the other hand we can identify the content of communication and the complexity of situated cognitions and actions as specific factors influencing the thinking pathways of students through interaction with a teacher. Out of this we conclude that interactions are successful only if the contents the students and the teacher 'talk about' (construct) are the same. Interactions are successful only if students and teacher are at the same level of complexity during interaction. Taking this into account a teacher who is trying to assess what the students already know about a subject, has to remember first what the students did last time, and has to start to ask questions at a low level of complexity for guiding them through the situation. If a teacher wants to give a hint during students' activities, he or she has first to observe the students' activities and to see at what complexity level they are. Then he or she can formulate hints at the same level. For guiding students towards a 'right way' through an experimental situation, the teacher should look where the students are in terms of content and complexity, then start there and go further. The students cannot understand principles or laws at the beginning of the learning process. As learners always interpret tasks on a complexity level they previously reached, the students are not able to solve tasks or problems which need a much higher level of complexity for their solution in the way the teacher intends. Moreover, the students will try to solve this problem, but 'only' at a lower level of complexity. If the teacher does not notice this phenomenon, misunderstandings can occur.

Finally, to answer the question formulated at the beginning of this paper: 'Why did the students not understand the teacher's explanation?', one answer could be that the content the teacher and the students thought (or spoke) about and/or complexity of the communication between teacher and students did not match.

References

Agre P. E. (1993). The symbolic worldview: Reply to Vera and Simon. *Cognitive Science,* 17, 61-69

Brown, J. S., Collins, A. & Duguid, P. (1989). Situated cognition and the culture of learning. *Educational Researcher,* 18 (1), 32-42

Clancey, W. (1993). Situated Action: A Neuropsychological Interpretation. Response to Vera and Simon. *Cognitive Science,* 17, 87-116

Damasio, A. R. (1994). *Descartes' Error. Emotion, Reason, and the Human Brain.* New York: Avon Books

de Jong, O. (1994). Studien über Fachlehrereinstellungen: mit oder ohne Kontext? (Studies about Science Teacher Attitudes) In Gramm et al. (Hrsg.). *Naturwissenschaftsdidaktik.* Westarp. Magdeburg

Greeno, J. G. (1991). Number sense as situated knowing in a conceptual domain. *Journal for Research in Mathematics Teaching,* 22, 170-218

Lang, M. (1998). *Bedeutungskonstruktionen und Lernen im Physikstudium.* (Development of Meaning and Learning in University Physics Courses) Dissertation. University of Bremen, Department of Physics and Electrical Engineering

Pfundt, H. & Duit, R. (1991). *Bibliography. Students' Alternative Frameworks and Science Education.* 3rd Edition. Kiel: IPN

Pöppel E. (1994). Temporal mechanisms in perception. *International Review of Neurobiology,* 37, 185-202

Rogoff, B. (1990). *Apprenticeship in thinking. Cognitive development in social context.* New York: Oxford University Press

Roth, W.-M. (1995). *Authentic School Science. Knowing and Learning in Open Inquiry Science Laboratories.* Dordrecht, Boston, London: Kluwer Academic Publishers

Schoster, A. (1998). *Bedeutungsentwicklungsprozesse beim Lösen algorithmischer Physikaufgaben, Eine Fallstudie zu Lernprozessen von Schülern im Physiknachhilfeunterricht während der Bearbeitung algorithmischer Physikaufgaben.* (Development of Meaning in Solving Algorithmic Physics Tasks) (Studien zum Physiklernen, Bd. 2). Berlin: Logos Verlag

Siegler, R. S. & Crowley, K. (1991). The microgenetic method. A direct means for studying cognitive development. *American Psychologist,* **46**(6), 606-620

von Aufschnaiter, S. & Welzel, M. (1996). Beschreibung von Lernprozessen. (Description of Learning Processes) In Duit, R. & v. Rhöneck, C. (Hrsg.). *Lernen in den naturwissenschaftlichen Fächern.* Kiel: IPN, pp. 301-327

von Aufschnaiter, S. & Welzel, M. (in press). Individual Learning Processes – A Research Program with focus on the complexity of Situated Cognition. *Proceedings of the 1st. European Conference of ESERA, September 2-6, 1997 in Rome*

Welzel, M. (1995). *Interaktionen und Physiklernen: Empirische Untersuchungen im Physikunterricht der Sekundarstufe I.* (Influence of Interactions in Learning Processes) Frankfurt a. Main; Bern, New York; Paris: Lang. (Didaktik und Naturwissenschaft; Bd.6)

Welzel, M. (1997). Investigations of Individual Learning Processes – a Research Program with its Theoretical Framework and Research Design. In Proceedings of the 3rd European Summerschool *Theory and Methodology of Research in Science Education.* Barcelona: E.S.E.R.A. and Universitat Autonoma de Barcelona, pp. 76-84

Welzel, M. (1998). The emergence of complex cognition during a unit on static electricity. *International Journal of Science Education.* **20** (9), 1107-1118.

Welzel, M. & Roth, W.-M. (1998). Do Interviews Really Assess Students' Knowledge? *International Journal of Science Education,* **20** (1), 25-44

Contact details for first authors of chapters in this book

Claudia von Aufschnaiter
Institute of Physics Education,
University of Bremen,
Box 330440,
28334 Bremen
Germany

Marta Gagliardi
Diparimento di Fisica
Viale B. Pichat 6/2
40127 Bologna
Italy

Miki Dvir
School of Education,
Faculty of Humanities,
Tel Aviv University,
Ramat Aviv,
Israel 69978

Olle Eskilsson
Kristianstad University
Box 59
MNA Institution
S-291 21 Kristianstad
Sweden

Contact details

Edgar Jenkins
Centre for Studies in Science and Mathematics Education,
University of Leeds,
Leeds
LS2 9JT
United Kingdom

Per Morten Kind
Norwegian University of Science and Technology
Dragvoll, Låven
N-7034 Trondheim
Norway

John Leach
Centre for Studies in Science and Mathematics Education,
University of Leeds,
Leeds
LS2 9JT
United Kingdom

Jean-François Le Maréchal
UMR GRIC-COAST
Université Lumière Lyon 2
5 avenue Pierre-Mendès-France
69676-BRON CEDEX 11,
France

Peter van Marion
Norwegian University of Science and Technology
Programme for Teacher Education
N-7491 Trondheim.
Norway

Robin Millar
Department of Educational Studies
University of York
Heslington
YORK YO1 5DD
United Kingdom

Contact details

Susan Molyneux-Hudgson
University of Bristol
Graduate School of Education
35 Berkeley Sq.
Bristol BS8 1JA
United Kingdom

Magnate Ntombela
Centre for the Advancement of Science and Mathematics Education
c/o University of Natal
King George V Avenue
Durban 4001
South Africa

Albert Chr. Paulsen
Royal Danish School of Educational Studies
Institute of Mathematics, Physics, Chemistry and Informatics
Emdrupvej 115B
DK-2400 Copenhagen NV
Denmark

Anja Schoster
Institute of Physics Education,
University of Bremen,
Box 330440,
28334 Bremen
Germany

Joan Solomon
The Open University
Walton Hall
Milton Keynes MK7 6AA
United Kingdom

Contact details

Pyotr Szybek
Lund University
Dept. of Education
Box 199
S-221 00 Lund
Sweden

Andrée Tiberghien
UMR GRIC-COAST
Université Lumière Lyon 2
5 avenue Pierre-Mendès-France
69676BRON CEDEX 11
France

J. Rod Watson
Kings College London
School of Education
Waterloo Road
London SE1 8WA
United Kingdom

Manuela Welzel
Institute of Physics Education,
University of Bremen,
Box 330440,
28334 Bremen
Germany